HVAC ESTIMATING

PROFESSIONAL REFERENCE

Adam Ding

Created exclusively
for DeWALT by:

PAL
publications

www.dewalt.com/guides

OTHER TITLES AVAILABLE

Trade Reference Series

- Construction
- Construction Estimating
- Construction Safety/OSHA
- Datacom
- Electric Motor
- Electrical
- Electrical Estimating
- HVAC/R – Master Edition
- Lighting & Maintenance
- Plumbing
- Plumbing Estimating
- Residential Remodeling & Repair
- Security, Sound & Video
- Spanish/English Construction Dictionary – Illustrated
- Wiring Diagrams

Exam and Certification Series

- Building Contractor's Licensing Exam Guide
- Electrical Licensing Exam Guide
- HVAC Technician Certification Exam Guide
- Plumbing Licensing Exam Guide

For a complete list of The DEWALT Professional Trade Reference Series visit **www.dewalt.com/guides**.

Pal Publications, Inc.
800 Heritage Drive, Suite 810
Pottstown, PA 19464-3810
800-246-2175

ISBN 978-0-9777183-5-1
11 10 09 08 07   5 4 3 2 1
Printed in Canada

A Note To Our Customers

We have manufactured this book to the highest quality standards possible. The cover is made of a flexible, durable, and water-resistant material able to withstand the toughest on-the-job conditions. We also utilize the Otabind process which allows this book to lay flatter than traditional paperback books that tend to snap shut while in use.

This Book Belongs To:

Name: _____

Company: _____

Title: _____

Department: _____

Company Address: _____

_____

_____

Company Phone: _____

Home Phone: _____

Preface

Is your HVAC business making the money you want it to make for you? Many HVAC contractors work hard with good craftsmanship and go the extra mile to satisfy customers, but they are not sure if they are making the most profit possible.

The only way to be sure is to take control of the numbers in your business. This book is written to introduce a down-to-earth HVAC estimating method. Instead of just discussing how to measure duct liners and count boilers, it provides you with various estimating tips, checklists, forms, data tables, etc. These contents are presented in a straight-forward, to-the-point, and "no-nonsense" format.

Best wishes,
Adam Ding

CONTENTS

CHAPTER 3 — *Pricing* 3-1

CHAPTER 4 — *Bidding* 4-1

CHAPTER 5 — *Post-Bid*. 5-1

CHAPTER 7 — *Technical Reference* . . 7-1

CHAPTER 8 — *Abbreviations* 8-1

CHAPTER 1
General Preparation

Estimating is the bloodline of contracting. It is here that profits are easily lost, even before a job begins. An estimator plays an essential role in helping a business succeed. Whether you are estimating for the business you own or for an employer, as an estimator you have a huge part in winning jobs and making them profitable. **Figure 1.1** shows an overview of the estimating process for HVAC contractors. So, how much do you know about your numbers?

GETTING ON THE BIDDER'S LIST

If you are on a general contractor's bidder's list, most likely you will receive notifications automatically for upcoming bids. The invitations could come in the form of fax, phone calls, e-mail, or mail. If you are not invited to bid for a while, you should contact the general contractors you work with to make sure you are still on their list.

Besides contractors bid lists, there are many other sources where you can look for business opportunities:

1. Advertisements in newspaper and trade journals
2. Bulletins posted in offices of government agencies, school districts, state universities, private colleges, etc.

3. News services of construction trade associations and public plan rooms
4. Direct invitations from owners or architects/engineers
5. Business contacts or word of mouth

Very often, HVAC contractors, especially new ones, have to team with a general contractor to win business. You will need to convince a good contractor that he should do business with you instead of someone else. You will have to convince him to put you on his bidder's list. This usually includes more work than just giving him a business card or letter of introduction. **Figure 1.2** provides a list of information you should provide for "pre-qualification."

A few points worth mentioning:

1. If you decide to visit a general contractor's office, make appointment in advance.
2. Be honest with all the information you provide.
3. Ask questions so you can later do a background check on him too.
4. Even if you are not a good fit to him, ask for a referral so you can look elsewhere.

Of course, you won't (and shouldn't) bid every job you are invited to. The essential question is: *how likely are you to get this job and make a profit?* If the answer is, "not very likely," then you'd better walk away instead of wasting time and effort on a meaningless bid.

GETTING ON THE BIDDER'S LIST*(cont.)*

Challenge yourself with the hard questions listed in **Figure 1.3**. Bidding a job could be one of the most nerve-wracking tasks you do on a regular basis. Know your strengths as well as limits, and do not be afraid to stand up to unreasonable demands from customers.

OBTAINING BIDDING DOCUMENTS

After deciding to bid a job, most HVAC contractors just make a request for drawings from the general contractors and wait for them to arrive. This is not necessarily the best way to get your drawings because you may not receive sufficient information to accurately bid the job. Bidding on incomplete information may cause you to bid too low or too high. If you bid high, the job could go to someone else. If you bid lower than what you should, you are left with two choices: not honoring your bid and seriously hurting the relationship with the general contractor, or absorbing the impact and hoping to make up for the loss later.

To have complete bid information, it is best to take a trip to the general contractor's office (or public plan rooms) where whole sets of documents are stored. Discuss the job with the general contractor's estimator to get a "general feel." Spend some time examining documents, and complete a general job information worksheet as shown in **Figure 1.4**.

1. Drawings (Architectural and Structural)

Flip through sheets such as life safety plan, floor plan, and wall section to get a general impression about the job. Find out the number of floors, square feet of floor space, and type of construction. Check the vertical distance between floors and locate mechanical rooms.

2. Drawings (Mechanical and Electrical)

Read relevant information shown on mechanical and electrical sheets. For example, an engineer may put a combined mechanical equipment schedule for both HVAC and Plumbing, which could show up on plumbing pages.

3. Specifications (General Requirements)

Read sections on instruction to bidders, general conditions, supplementary and special conditions, completion dates, payment schedule, contract clauses, allowances, alternates, etc.

4. Specifications (Mechanical and Electrical)

Read through sections on general mechanical requirements, coordination with other trades such as plumbing and electrical, etc.

At the end of the visit, ask if the general contractor is willing to give you relevant drawings and specs. If they are, fill out a document request worksheet (**Figure 1.5**) and give it to their estimator. Very often, you will have to buy them at your own expense. In any case, make sure you have access to complete information required for take-off and pricing.

PREPARING A PRELIMINARY ESTIMATE

Whenever possible, always prepare a detailed and accurate estimate. Resist the "budget estimates" requests from general contractors, as those are seldom real projects.

Sometimes general contractors negotiate work with the owner. Or occasionally, the owner may offer you the direct contract to do the HVAC work without involving a general contractor. In these cases, you will be required to estimate HVAC costs based on preliminary information, often nothing more than a set of architectural drawings.

These projects are often called "design-build" jobs, as opposed to "plans-specs" jobs where complete mechanical drawings and specs exist. A preliminary estimate can be prepared from experience or historical data. There are several ways to price this type of job:

1. Apply unit prices to number of functional units (e.g. how many residential condos, hotel rooms, school students, hospital beds, etc.)
2. Apply unit prices to different functional areas (e.g. areas for parking, residential living, commercial retail, or common facilities)
3. Apply unit prices to different systems or components (see **Figure 1.6**)

Preliminary estimates should never be used to bid "plans-specs" jobs because this method is not accurate enough. They are useful for getting a rough idea of the total costs, but costs should be verified later by more detailed estimates.

COMPUTERIZED ESTIMATING AND BIDDING

Today a lot of folks are using computers to do estimating. Studies show that computerized estimating can save 75% of time, compared to estimating manually with pencil and paper. Many small companies, where the owner does the estimating, can use computerized estimating to free more time for field management and business marketing.

Spreadsheet programs have been widely used for estimating and other functions, but they require an understanding of how formulas work. You might have a hard time making them work to fit your specific needs. When calculations gets complicated, writing correct formulas becomes more difficult.

Dozens of commercial estimating software packages are available today to help with HVAC estimating and bidding. They are easier to use, can perform calculations automatically, and many have a pricing database built in them. A question list shown in **Figure 1.7** can help you make a wise purchase decision.

FIGURE 1.1 — HVAC ESTIMATING PROCESS

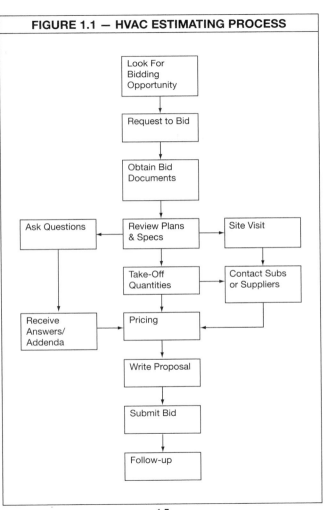

FIGURE 1.2 — WHAT TO INCLUDE ON A MARKETING LETTER

- Company name, address, telephone number, fax number, and email addresses (if applicable)
- The companies status as a corporation, partnership, or sole proprietorship
- Years in business under the current company name
- Names of company officers, principals, partners, or owners
- The types and size of projects you would like to bid
- Type of work you do with your own forces
- Type of work you subcontract to others
- List of jurisdictions in which your firm is licensed
- Average annual dollar volume for the past three years
- Number of current employees in the office and field
- List of current jobs including contract amount
- List of completed projects including contract amount, completion date, architect name and general contractor name, contact, and phone number
- Three or more business references including company, contact, and phone number
- Ability to bond and bonding capacity, bonding agent name, contact, and phone number
- Insurance agent name, contact, phone number, and sample certificate of insurance
- Bank name and contact
- A list of trade associations of which you are a member

FIGURE 1.3 — SHOULD YOU BID THIS JOB?

- Is this job similar to the ones you normally do (type, size, complexity, quality, etc.)?

- Do you understand the scope of the work?

- Do you know the general contractor?

- Do they have a bad reputation like non-payments or bid shopping, etc.?

- Does this general contractor coordinate subs well?

- Is the general contractor using you to do a price check?

- Is this a real job or just a budgeting exercise?

- Are there many HVAC contractors bidding the job? Who are they?

- Have you bid against these HVAC contractors before? Did you win?

- How far is this job from your office?

- Is the job in "your territory"? Are you licensed to do work in the area?

- Do you know the material suppliers in that area as well as local building codes?

- Is the job in a heavy union area? Is your company union or non-union?

- How's owner's reputation? Does he have enough money to finance the job?

- How and when you will expect to receive payment once the job starts?

FIGURE 1.3 — SHOULD YOU BID THIS JOB? *(cont.)*

- What are the procedures for change orders (if allowed)? How long will you have to wait before you get paid for changes?

- Who is the architect or engineer? How is his reputation?

- Are drawings complete or is the job still in design phase?

- How many jobs do you have under way? If another job is awarded, will your crew be available to do it?

- Do you have to invest in special tools and equipment to do the work?

- Is there anything unusual about the job such as night shifts or limited working hours?

- Overall, can you expect a decent profit if you get the job?

FIGURE 1.4 — GENERAL JOB INFORMATION WORKSHEET

Job Name: _____

Ref No. _____   Checked By: _____

Bid Due Date/Time: _____

Job Location: _____   Tax Rate: _____

Construction Type: ❑ New Construction   ❑ Renovation

Gross Floor Area: _____   Number of Floors: _____

No. of Condo Units: _____   No. of Hospital Beds: _____

No. of Hotel Rooms: _____

Send Bid To: ❑ Owner:_____
   ❑ General Contractor: _____

Owner: _____   Tel No.: _____

General Contractor: _____   Tel No.: _____

Architect: _____   Tel No.: _____

Mechanical Engineer: _____   Tel No.: _____

Plans/Specs or Design/Build: _____

CAD required? _____

Project Start Date: _____   Project Finish Date: _____

Retainage: _____   Liquidated Damages: _____

Warranty Period: _____

Labor Conditions: ❑ Prevailing Wages ❑ Union ❑ Open Shop

Bond Required: ❑ Yes   ❑ No

Test and Balance ❑ Include   ❑ Exclude

Send RFI to   ❑ General Contractor   ❑ Mechanical Engineer
   ❑ Other: _____

Work Done by Owner's Contractor: _____

Cost Breakdown Requirements: _____

Allowances: _____   Alternates: _____

FIGURE 1.5 — DOCUMENT REQUEST WORKSHEET

Job Name: _____

Ref No. _____ By: _____

To (General Contractor/Owner's Rep/Blueprint Shop): _____

We need the following documents in order to bid the above project:

Drawings	Sheets/Sections
Specifications	
HVAC	
Plumbing	
Electrical	
Fire Protection	
Civil	
Landscape/Irrigation	
Architectural	
Interior Design	
Structural	

Please deliver the above documents to:

Contact Person: _____

Company Name: _____

Delivery Address: _____

Telephone Number: _____

Thank You.

FIGURE 1.6 PREPARING A PRELIMINARY ESTIMATE

Note: All numbers and dollar amounts in these examples are fictional and only for illustrative purpose.

By Function Unit:
HVAC costs for a hotel of 110 rooms: 110 × $7,000 = $770,000

By Function Area:
In a high-rise condominium:
Parking/Retail area: 60,000 Sq. Ft. × $3.20 = $192,000
Residential suites area: 120,000 Sq. Ft. × $18.35 = $2,202,000
Total HVAC costs: $192,000 + $2,202,000 = $2,394,000

By System Component:

Items	QTY	Unit	Unit Price	Subtotal
Air Handlers and Fans				
central heat and cooling air-handling units	1	no	30,000.00	$30,000
return flow-units – heat-recovery	1	no	36,000.00	$36,000
crawlspace humidifiers	1	ls	N/A	$0
air-cooled chiller unit	1	no	27,500.00	$27,500
cooling tower	1	ls	N/A	$0

1-13

FIGURE 1.6 PREPARING A PRELIMINARY ESTIMATE (cont.)

Items	QTY	Unit	Unit Price	Subtotal
associated circulation pumps	2	no	2,500.00	$5,000
expansion tank	1	no	1,500.00	$1,500
air-separator	1	no	1,500.00	$1,500
chemical treatment, etc.	1	ls	3,500.00	$3,500
re-heat coils	6	no	950.00	$5,700
miscellaneous return and exhaust fans	1	no	7,000.00	$7,000
washroom / bathing area exhausts	1	ls	N/A	$0
range exhausts (excludes hoods)	1	no	750.00	$750
corridor supply ductwork and air distribution	730	m2	60.00	$43,800
Miscellaneous				
testing, balancing, and commissioning	1	ls	10,000.00	$10,000

FIGURE 1.6 PREPARING A PRELIMINARY ESTIMATE (cont.)

Items	QTY	Unit	Unit Price	Subtotal
Piping				
natural gas	25	m	65.00	$1,625
natural gas shut-off valves	2	no	500.00	$1,000
heating and chilled supply/return mains pipe work	310	m	75.00	$23,250
connecting pipe work to existing mech. room	1	ls	20,000.00	$20,000
condensate pipe work	25	m	45.00	$1,125
Heating				
gas-fired mid-efficiency heating boiler	1	no	22,000.00	$22,000
associated circulation pumps	2	no	2,500.00	$5,000
breaching, chemical treatment, etc.	1	ls	3,500.00	$3,500
expansion tank	1	no	1,500.00	$1,500
air-separator	1	no	1,500.00	$1,500
resident rooms hot-water radiant panel heaters	33	no	750.00	$24,750
hot-water heaters – other areas	4	no	750.00	$3,000

1-15

FIGURE 1.6 PREPARING A PRELIMINARY ESTIMATE (cont.)

Items	QTY	Unit	Unit Price	Subtotal
DDC controls to equipment				
new computer software and hardware	1	ls	10,000.00	$10,000
central heat and ventilation units – multi-zone	1	no	10,000.00	$10,000
return flow-units – heat-recovery	1	no	5,000.00	$5,000
exhaust fans and re-heat coils	7	no	950.00	$6,650
crawlspace humidifiers	1	no	N/A	$0
heating and cooling plant	1	ls	15,000.00	$15,000
circulation pumps	5	no	350.00	$1,750
resident room thermostats	15	no	350.00	$5,250
other thermostats	4	no	950.00	$3,800
miscellaneous control dampers	4	no	550.00	$2,200
motorized dampers	1	ls	Included	$0
Total HVAC Costs				**$340,150**

FIGURE 1.7 – ESTIMATING SOFTWARE EVALUATION

❏ Is the software simple to use and easy to understand?

❏ How does the software work for estimating?

❏ Is it a simple spreadsheet application or specialized software with database?

❏ Does it do what you need it to do?

❏ What is required to maintain the software (e.g. digitizers for quantity take-off)?

❏ How does the software maintain and update material and labor prices?

❏ Does the software have features like a supplier and subcontractor contact database?

❏ How well does the software present the information required (e.g. automatically creating an estimating summary report)?

❏ Can the software help with the bid preparation and submission?

❏ Does the software have additional functions like bookkeeping, job costing, project document management, service recording, and billing?

❏ How much time can you expect to save by using the software?

❏ Is it the latest technology or being phased out?

❏ What is the upfront purchase cost for the software?

❏ Is an upgrade available and at what cost?

FIGURE 1.7 — ESTIMATING
SOFTWARE EVALUATION *(cont.)*

❏ Do you have to pay for periodical license renewal and upgrades?

❏ Can you try the software for a period of time without purchasing?

❏ Is training required? Does the software vendor provide training and technical support for his software? How much does he charge?

❏ How long has the software vendor been in the business? How many customers does the vendor have? What other software products does the vendor sell?

CHAPTER 2
Quantity Take-Off

A good or bad estimate all starts with the takeoff. You need to get used to a set of routine procedures and apply them to estimating every job. By doing so, nothing will be left out and you will save time and effort in the long run.

REVIEWING DOCUMENTS

The best way to learn estimating is just doing it. But, that does not mean you can jump into taking off quantities immediately. Spending sufficient time reviewing documents can reduce errors and omissions. Remember, better slow than sorry.

1. Check Documents

Once you receive a set of bid documents, open the package and check to see if any information required is missing (eg. the mechanical specifications may have been left out, or some drawing sheets may be missing). Contact the sender to ask for the missing pieces.

If the package seems to be "complete," then flip through pages to see what qualities these documents have. Sometimes drawings are so schematic that you may not want to bid the job any more. Without going into specifics, the worksheet shown in **Figure 2.1** can help you make an evaluation.

2. Study Drawings

Thoroughly study drawings sheet by sheet until they are fully understood. Go over each floor tracing heating/cooling equipment and ductwork. Visualize how components fit into the job and mark them with different colored pencils. Carefully read and mark all special notes on plans. Note important details such as:

- Scale of drawings
- Height of structure for vertical lifts, cranes, etc.
- Equipment room dimensions and air unit space limitations for agreement with selected equipment
- Electrical characteristics
- Elevations for scaffold and ladder work

After studying the drawings, you should have a clear mental picture of the job, understanding how complex it is, how long it might take to finish the job, and anything special needed to install the required systems.

3. Read Specs

Read specifications, paragraph by paragraph, at least twice. Specs can be complex, consisting of a whole division with many pages of detailed information. Or they can be simple, laid out on the same page with the drawings.

Make notes in the following aspects of specifications:

- Scope of work (what is to be included)
- Expensive and special items
- Manufacturer, type, and size of equipment
- Guarantee, service period, instructions, test and balance, etc.
- Errors and omissions

4. Deal with Conflicts

After reviewing drawings and specifications individually, you can verify if the two are compatible. If they conflict with each other, normally specifications overrule.

However, when drawings and/or specs conflict with the governing mechanical code, the code always overrules. Remember, *the code is the final word*.

If you do not feel comfortable about any issue, get clarification from the design engineer.

5. Summarize Scope

Use the worksheet in **Figure 2.2** to make notes. Write down as much as you can, including types and sizes of equipment, duct sizes, material, etc. Make reference to related drawing sheets and spec sections. The completed worksheet is an ideal template to define the job scope. Try your best to include everything. It is much better to be low in certain items than to leave them out entirely.

For unclear details, you should ask the designer instead of making assumptions. Questions should be grouped as much as possible since calling every ten minutes with a new question is not considerate. Use an RFI form (**Figure 2.3**) to send questions. If you do not get an answer after a few days, follow up with a phone call. Use a phone memo form (**Figure 2.4**) to record the conversation.

Figure 2.5 lists some common design features that will add extra costs compared to regular HVAC systems.

VISITING THE SITE

Before you bid a job, you should always visit the site. Very often, site conditions are different from what you see on the mechanical drawings. A site visit will enable you to make additional allowances to account for items that are hard to quantify.

The following is a list of things you should take note of when visiting the site:

- **Site Access**: road conditions; weight and height limits of bridges; ease of delivering tools, material, and equipment; site security; etc.
- **Existing Building**: Demolition or renovation requirements, hazardous materials, operation while construction, working hour restrictions, dust and noise reduction, etc. Obtain as-built drawings for the existing facility and try to contact the mechanical contractor who built it.

- **Unusual Job Conditions**: Watch for situations such as high ceilings, flooded basements, crawl spaces, etc. Identify future locations for trash dumpsters, material storage trailers, portable toilets, etc.

You can use the Site Visit Worksheet in **Figure 2.6** as a general guide. Note that sometimes repeated site visits might be necessary to get more information. It is recommended to start each estimate within 24 hours of site inspection.

TAKING OFF QUANTITIES

To "take-off" a job means that you "take" the information "off" the documents and translate it into a list of items with quantities. It is important to decide how you should break out numbers. Sometimes owners put their detailed cost breakdown requirements in the specifications, which you must follow. Otherwise, depending on each job, you can break out your take-off by systems, floors, phases, base price, alternates, etc.

The actual take-off can be described in 3 easy steps:

A. Measure each item

General guidelines are as follows:

1. Do not scale drawings. Instead, try to use the dimensions specified.
2. Each system should be taken off separately. Apply color coding to differentiate.

3. Identify major supply and return lines and quickly trace them out.

4. Note the number of similar spaces. Repetitive details like the same types of retail bays on a commercial job can save time in actual quantity take-off.

5. Beware of reduced size drawings, as scale noted on them is often wrong.

6. Use a calculator that prints on paper ("cashier" type).

In reality, because you may never be able to complete a quantity take-off without interruptions, marking drawings is important to remind you of what has been taken-off. Unmarked drawings could have you double counting or missing things.

Remember to include required items that may not appear on the drawings such as fitting, hangers, shields, inserts, bolts, nuts, gaskets, etc. **Figure 2.7** provides a HVAC Quantity Take-off Checklist for your reference.

B. Record quantities

Because there is so much information to be taken off the drawings, it is best to use some kind of forms to stay organized. Good estimating forms provide a permanent record and reduce the work load and chance of error. **Figure 2.8** through **Figure 2.12** give standard quantity take-off forms you can customize and use.

While using these forms:

- **Keep different system items separate**. Don't combine similar items for different systems, even if the materials are the same type.
- **Make detailed reference**: Note which sheet you find items on (drawing number, detail number, grid reference, etc.); where they exist in the building (floor level, room number, etc.); and what details these items have. For example, the word "pump" without any other description is not clear enough if there are several pumps in the job.

C. Summarize Quantities

When you are done with take-off, calculate total quantities at the bottom of each worksheet. Round up the numbers, according to the precision desired (e.g. you might just want duct lengths to be whole numbers without decimal).

Then, add the quantities on each sheet and group them for the same type of items. The results can be used later for material quote requests and pricing.

CONTACTING SUBCONTRACTORS

Although an HVAC contractor is a subcontractor to the general contractor of the job, the HVAC contractor may have his own subs to do the things he normally does not do. Sometimes it can be cheaper or faster to get someone else to do portions of the work.

Subcontractors might include (depending on scope):

- Insulation
- Sheet Metal
- Automatic Controls
- System Test and Balance
- Concrete
- Paint

If you need subs to bid a job, it is best to notify them as early as possible. Go over the job with them and provide relevant information (maybe an extra copy of drawings and specs). If delayed, the subs either can not finish the bid or just bid high to make sure they cover their costs. This could make your total bid too high, and you could lose the deal.

FIGURE 2.1 — CHECKING DOCUMENT COMPLETENESS

Job Name: _____

Ref No.: _____   Checked By: _____

HVAC Plan

❑ Drawn to the same scale as the architectural plans
❑ Separate from the plumbing plans and electrical plans
❑ Shows partitions, room layouts, and fire and smoke rated partitions
❑ Shows ductwork layout including size and pressure class
❑ Includes all necessary details, sections, schedules, and notes to show the extent of the work
❑ Includes building heating and cooling loads (in BTU's/Hr), temperature differentials used, and rated capacity of heating units
❑ Shows location of fire and smoke dampers
❑ Shows coil and tube pull areas
❑ Indicates required code clearance areas
❑ Includes all devices such as balancing dampers, splitter dampers, volume extractors, balancing valves, thermometers, pressure gauges, instrument-flow fittings, and instrument-access panels required for balancing

Specifications

❑ Basic mechanical materials and methods
❑ Testing, adjusting, and balancing
❑ Manufacturers' names, products brands, or catalog numbers
❑ Required performance criteria for all materials and assemblies
❑ Installation procedures, coordination procedures, and clean up methods
❑ Operations and maintenance manuals covering each item of equipment

Conclusions:

Do you think these documents are complete enough for bidding?
❑ Yes   ❑ No

Issues to be resolved with the design engineer:

_____
_____

FIGURE 2.2 — HVAC SCOPE WORKSHEET

Job Name: _____

Ref No.: _____   Checked By: _____

EQUIPMENT

Heat Generation

❑ Boiler: _____

     ○ Gas Fired: _____

     ○ Oil Fired: _____

     ○ Electric: _____

     ○ Hot Water: _____

     ○ Steam:_____

     ○ Low Pressure: _____

     ○ High Pressure: _____

     ○ Type: _____

     ○ Horsepower/MBH: _____

❑ Boiler Circulation Pumps: _____

❑ Expansion Tanks: _____

❑ Breeching: _____

❑ Stacks: _____

❑ Blow Down Tank: _____

❑ De-aerator:_____

❑ Boiler Chemical Treatment: _____

❑ Water Softener for Boilers: _____

Electric Heat

❑ Baseboard Heaters: _____

❑ Wall Heaters: _____

❑ Unit Heaters: _____

Refrigeration

❑ Centrifugal Chiller: _____

❑ Heat Recovery Chiller: _____

❑ Water Cooled Chiller: _____

FIGURE 2.2 — HVAC SCOPE WORKSHEET (cont.)

EQUIPMENT (cont.)

❑ Air Cooled Chiller: _____

❑ Condensers: _____

❑ Condensing Units: _____

❑ Cooling Towers: _____

❑ Expansion Tanks: _____

❑ Chilled Water Pumps: _____

❑ Condenser Water Pumps: _____

❑ Plate Type Heat Exchangers: _____

❑ Variable Speed Drives: _____

❑ Ice Storage Tanks: _____

❑ Water Treatment: _____

Liquid Heat Transfer (Heating)

❑ Steam to Steam Heat Exchangers: _____

❑ Steam to Water Heat Exchangers: _____

❑ Hot Water Generators: _____

❑ Primary Heating Pumps: _____

❑ Secondary Heating Pumps: _____

❑ Coiling Circulating Pumps: _____

Steam & Condensate

❑ Steam PRV Stations: _____

❑ Condensate Pumps –Simplex: _____

❑ Condensate Pumps –Duplex: _____

❑ Steam Humidifiers: _____

Liquid Heat Transfer (Cooling)

❑ Circulating Pumps: _____

❑ Coil Circulating Pumps: _____

❑ Fan Coil Units: _____

❑ Heat Pumps: _____

❑ Computer Room A/C Units: _____

FIGURE 2.2 — HVAC SCOPE WORKSHEET (cont.)

EQUIPMENT (cont.)

❏ Thru-wall A/C Units: _____

❏ Window A/C Units: _____

Energy Recovery

❏ Air to Air Heat Exchanger: _____

❏ Heat Recovery Wheel: _____

❏ Glycol Coils: _____

❏ Glycol Pumps: _____

Air Handling Equipment

❏ Pre-engineered Air Handling Units: _____

❏ Custom Built Air Handling Units: _____

❏ Roof Top Air Handling Units:_____

Exhaust & Ventilation

❏ Bathroom Exhaust Fans: _____

❏ Kitchen Exhaust Fans: _____

❏ Ecology Air Handling Units:_____

❏ Smoke Exhaust Fans:_____

❏ Stairwell Pressurization Fans: _____

❏ Car Park Exhaust Fans: _____

❏ Fume Hood Exhaust Fans: _____

Fuel Oil System

❏ Oil Storage Tank:_____

❏ Propane Storage Tank:_____

❏ Oil Pump Sets:_____

Other Equipment

❏ Motor Starter & MCC's: _____

DUCTWORK

❏ Low Pressure Ductwork: _____

❏ Medium Pressure Ductwork:_____

❏ High Pressure Ductwork:_____

FIGURE 2.2 — HVAC SCOPE WORKSHEET *(cont.)*

❑ Pre-Fabricated Round Duct: _____

❑ Pre-Fabricated Round Double Wall Duct: _____

❑ Sanitary Exhaust Duct: _____

❑ Specialty Duct Systems: _____

    ○ Stainless Steel: _____

    ○ Aluminum: _____

    ○ Acid Resistant: _____

    ○ Flat Oval: _____

    ○ Black Iron: _____

    ○ Pre-Manufactured Plenums: _____

❑ Duct Lining: _____

❑ Duct Insulation: _____

❑ Fire Dampers: _____

❑ Balancing Dampers: _____

❑ Sound Attenuators: _____

❑ Automatic Dampers: _____

❑ Outside Louvers: _____

PIPING

❑ High Pressure Steam: _____

❑ Medium Pressure Steam: _____

❑ Low Pressure Steam: _____

❑ Condensate: _____

❑ Hot Water: _____

❑ Chilled Water: _____

❑ Condenser Water: _____

❑ Fuel Oil System: _____

❑ Pre-insulated Piping System: _____

❑ Fan Coil Drains: _____

❑ Glycol: _____

❑ Snow Melting: _____

FIGURE 2.2 — HVAC SCOPE WORKSHEET *(cont.)*

PIPING (cont.)

❑ Insulation: _____

❑ Other: _____

DUCTWORK TERMINAL DEVICES

❑ Variable Air Volume Boxes: _____

❑ Constant Air Volume Boxes: _____

❑ Fan Powered Terminal Boxes: _____

❑ Diffusers:_____

❑ Grilles: _____

❑ Laminar Flow Diffusers: _____

❑ Terminal Humidifiers: _____

❑ HEPA Diffusers: _____

❑ Other: _____

PIPING TERMINAL DEVICES

❑ Wall Fin Elements:_____

❑ Wall Fin Enclosures: _____

❑ Runtal Type Heaters: _____

❑ Radiant Heat Panels : _____

❑ Convectors: _____

❑ Radiators: _____

❑ Cabinet Heaters: _____

❑ Unit Heaters: _____

❑ Reheat Coils: _____

❑ Unit Ventilators: _____

❑ Induction Units: _____

❑ Air Curtains: _____

❑ Gas Fired Unit Heaters: _____

❑ Infra-Red Heaters:_____

❑ Radiant Floors: _____

❑ Other: _____

FIGURE 2.2 — HVAC SCOPE WORKSHEET (cont.)

CONTROLS

Central Equipment

❏ Air Compressors: _____

❏ Computer Hardware: _____

❏ Computer Software: _____

❏ Computer Programming: _____

❏ Other: _____

Control Points

❏ Pneumatic Control Points : _____

❏ DDC Control Points: _____

❏ Radiation Valves: _____

❏ VAV Actuators: _____

❏ Supply and Exhaust Air Valves: _____

❏ Fume Hood Control Modules: _____

❏ Damp Actuators: _____

❏ Thermostats: _____

❏ Monitoring, Sensing, and Measuring Devices: _____

❏ Control Valves: _____

❏ Other: _____

MISCELLANEOUS

❏ Electrical Wiring: _____

❏ Testing, Adjusting, Balancing: _____

❏ Fastening: _____

❏ Foundations and Support: _____

❏ Excavation and Trenching: _____

❏ Rigging: _____

❏ Painting: _____

❏ Tools and Special Equipment: _____

FIGURE 2.3 — RFI FORM

Request for Information

From:	RFI#:
	Date:
	Job Name:
	Job Ref No:

To:

Attn:

We are requesting that you review the following matter and advise on how we are to proceed. In the event that your determination constitutes a change to the contract documents, please issue an addendum if possible.

Issue:

Please do not hesitate to contact us if you have any questions.

Signature:

Enclosures:

Reply:

Signature:

Date:

FIGURE 2.4 — PHONE MEMO WORKSHEET

Confirmation of Verbal Communication

To:	Date:
Address:	Time:
City:	
State:	Zip:
Attn:	
Subject:	
Project:	
This memo confirms the conversation:	
Between And	
On (date)	
Brief summary of conversation:	

Please notify our office at once if you do not concur with this conversation summary.

FIGURE 2.5 — COST IMPLICATIONS OF MORE EXPENSIVE DESIGN FEATURES

Design Features	Implications
100 percent outside air units	Designed to "pre-condition" the outside air before it enters the building
Absorption or gas-operated chillers	More expensive than normal electrical-powered chillers
Air valves	Precision air metering devices found in laboratories
Commissioning	Depending on scope
Computer room units	Specialty air conditioning systems containing their own controls
Custom size units	Units over 25,000 cfm in capacity
Dehumidification/ Humidification	Additional control components and systems
Energy recovery systems	Significant initial investment in construction
Geo-thermal system	Underground work required
Ice storage	Require the use of anti-freeze and careful equipment selection
Indoor Air Quality testing	Depending on scope
Sound level requirements	Use of sound traps, vibration isolation hangers, etc.
Stainless steel for cooling tower, piping, or ductwork	Uncommon materials
Using hot water to heat the air	More HVAC costs compared with electrical heating

FIGURE 2.6 — SITE VISIT WORKSHEET

Job Name: _____

Ref No.: _____

Visited By: _____

Date: _____

Address: _____

Directions: _____

Site Access: _____

Site Description: _____

Hazardous Material: _____

Existing Building Maintenance Conditions: _____

Demolition/Relocation/Upgrade
of Existing HVAC System: _____

Neighborhood Information: _____

Local Mechanical Codes: _____

Local Supplier: _____

Labor Conditions: _____

Equipment Required: _____

Dust control and noise abatement requirements: _____

Comments: _____

Photos attached: _____

FIGURE 2.7 — HVAC QUANTITY
TAKE-OFF CHECKLIST

Equipment

Boilers, burners, furnaces (EA)

Unit heaters, convectors (EA)

Compressor or chiller units (EA)

Condensers (EA)

Receivers (EA)

Cooling towers (EA)

Chimneys (EA)

Heat exchangers (EA)

Air handling units (EA)

Exhaust fans (EA)

Ventilators (EA)

Expansion tanks (EA)

Storage tanks (EA)

Heat pumps (EA)

Condensate pumps (EA)

Dust collector (EA)

Fume hoods (EA)

Piping

Chilled water piping (LF)

Hot water piping (LF)

Condenser water piping (LF)

Refrigerant lines piping (LF)

Steam piping (LF)

Condensate piping (LF)

Oil piping (LF)

Gas piping (LF)

Pipe insulation (LF)

Fittings (EA)

EA = Each, LF = Linear Foot, L/S = Lump Sum, SF = Square Foot

FIGURE 2.7 — HVAC QUANTITY
TAKE-OFF CHECKLIST *(cont.)*

Valves (EA)
Pipe painting (L/S or SF)

Ductwork
Supply ducts (LF, SF, or Pounds)
Return ducts (LF, SF, or Pounds)
Ductwork insulation (SF)
Louvers, diffusers, registers, dampers, and grilles (EA)
Fittings (EA)
Valves (EA)
Filters (EA)

Miscellaneous
Packaged control (EA)
Starters (EA)
Motors (EA)
Thermostat (EA)
Humidistat (EA)
Air purification (L/S)
Humidification and Dehumidification (L/S)
Test and balance (L/S)
Refrigerant (Lbs)
Vibration isolation (L/S)
Access panels (EA)
Valve tags (EA)
Covers and frames (EA)
Rigging (L/S)
Equipment room construction (L/S)
Concrete pad (EA or SF)
Excavation and backfill (L/S)
Electrical wiring (L/S)

EA = Each, LF = Linear Foot, L/S = Lump Sum, SF = Square Foot

FIGURE 2.8 — GENERAL QUANTITY TAKE-OFF WORKSHEET

Job Name:		Ref No.:		Estimator:	
Estimate Date:			Sheet Numbers:		
Item Description	**Details**				**Extension**
	Length	Width	Height	Count	
Total					

FIGURE 2.9 — HVAC EQUIPMENT QUANTITY TAKE-OFF WORKSHEET

Job Name:	Ref No.:	Estimator:	
Estimate Date:		Sheet Numbers:	

System	Item	Capacities/ Dimensions	Quantity
Total			

FIGURE 2.10 — HVAC PIPING QUANTITY TAKE-OFF WORKSHEET

Job Name: _____   Ref No.: _____   Estimator: _____

Estimate Date: _____   Sheet Numbers: _____

System	Pipe Diameter																Misc.
	1/8"	1/4"	3/8"	1/2"	3/4"	1"	1 1/4"	1 1/2"	2"	2 1/2"	3"	4"	5"	6"	8"	10"	12"	15"	
Total																			

FIGURE 2.11 — HVAC DUCTWORK QUANTITY TAKE-OFF WORKSHEET

Job Name:

Ref No.:

Estimator:

Estimate Date:

Sheet Numbers:

Gauge	Size	Lining	Insulation	Length	Total	Lbs/Ft	Total Lbs
Total							

FIGURE 2.12 — HVAC ACCESSORIES QUANTITY TAKE-OFF WORKSHEET

Job Name:

Ref No.:

Estimator:

Estimate Date:

Sheet Numbers:

Items	Dimension/Description	Quantity
Total		

CHAPTER 3
Pricing

Pricing is the process you use to convert the quantities you took off into dollar values. You should NOT start pricing until you have completed your take-off for all HVAC items required in the job scope.

Generally, pricing can be done in these steps:

1. Summarize the material quantities you have taken-off. Combine the numbers for the same items and allow for reasonable waste.
2. Price materials based on the quotes you received from the suppliers.
3. Price labor based on the man-hour information.
4. Add indirect costs such as overhead, bond, insurance, permits, etc.
5. Add your desired profit to get a total price.

An important thing to keep in mind is that you are pricing labor and material according to the time when the work is expected to be done, not when the job is being estimated. Most HVAC work is done several months after the bid is submitted. There is no way to be sure what prices will be in three to six months. Therefore, if you expect there will be a price escalation or labor shortage in the future, it is best to make some allowances now or get price guarantees in writing from your suppliers and subcontractors.

PRICING MATERIAL

Basic Formula:

Material Price = Quantity × Material Unit Price

Get Quotes

Always request quotes from at least three suppliers. Provide a complete list of the items. Specify as much information as possible including product type, model number and make, quantity, etc. Sometimes for special items, attach copies of details from drawings.

Some suppliers might offer to figure the total material prices for you. Check their figures carefully by reading their quotes carefully and verifying the following:

- Unit Price
- Delivery Charge
- Sales Tax (payable to all levels of governments: city, county, state, and federal).
- Minimum Order Quantity
- Expected Price Escalation (i.e. the prices by the time you actually place order)
- Discount Rate

It is better to prepare your own material estimate than to have suppliers figure it for you. **Figure 3.1** is a sample material pricing sheet you can use. **Figure 3.2** explains some terms related to material shipping.

Material Discounts

A discount is a deduction figured in percentage from the original price offered by the material supplier. The base is the original price of the item, or "list price." The price after discount is called "net price."

PRICING MATERIAL *(cont.)*

The amount of discount is the difference between these two prices.

A. Estimating a Single Discount
Formula
Net Price = List Price $\times$ (1 − Discount Rate)

Estimating Example:
A valve is listed at $15.00 and the supplier offered a discount of 25%, then:

The net price is $15.00 $\times$ (1 − 25%) = $11.25

The amount of discount is $15.00 − $11.25 = $3.75

B. Estimating Multiple Discounts
Discounts are often offered in multiple terms, e.g. 25-10%, 50-10-15%, etc. The meaning of such a discount as 25-10%, is that 25% will first be deducted from the list price and then 10% will be deducted from the remainder.

Modified Formula:
Net Price = List Price $\times$ (1 − Discount Rate 1) $\times$ (1 − Discount Rate 2) $\times$ ......

Estimating Example:
A valve is listed at $15.00 and the supplier offered a discount of 25-10%, then:

The net price after first discount is
$15.00 $\times$ (1 − 25%) = $11.25

The net price after second discount is
$11.25 $\times$ (1 − 10%) = $10.13

The amount of discount is $15.00 − $10.13 = $4.87

C. Simplified Calculation

For multiple discounts, sometimes there are too many numbers to crunch, like a rate of 40-10-10-5%. Therefore, it is more convenient to apply a "resulting rate" for a certain discount. Let's use the above examples again.

For a single discount rate of 25%, you can just apply a "resulting rate" of 0.75.

The net price for that valve is:
$15.00 × 0.75 = $11.25

For a double discount rate of 25-10%, you can just apply a "resulting rate" of 0.675.

The net price for that valve is:
$15.00 × 0.675 = $10.13

A data table is given in **Figure 3.3** to simplify the calculation on multiple discounts.

For example, for a discount rate of 40-10-10-5%, the "resulting rate" is 0.4617.

The net price for that valve is:
$15.00 × 0.4617 = $6.93.

No more math headaches!

PRICING LABOR

Basic Formula:

Labor Price = Quantity × Man-hour per Item ×
Labor Hourly Wage

There are quite a few ways to price labor. For example, some HVAC contractors just apply a labor unit price rate (i.e. how much it costs to install each item) to their quantities.

Sometimes it is hard to track unit prices, especially when they change very often. What works on one job may not work well on the other. If you use the same money-making unit prices on all jobs, you may lose money. You may have a "gut feeling" that things are a little different from your last job, but you may not know how to compensate for those differences.

A better way is to use man-hours (i.e. what a man can do within an hour). This information tends to remain relatively stable from job to job. For example, from previous jobs you have done, you know that it takes you about 3 hours to install a certain type of boiler. You are paid $35.00 per hour. Then the labor to install 2 EA of boilers will cost:

2 boilers × 3 man-hours × $35.00/man-hour = $210

A common mistake is to forget that man-hours vary for installation conditions. This book does give man-hours for some common HVAC items in **Chapter 6**, but you should not use the information without adjusting it for specific job situations. Examples include:

- Job Size
- Overtime
- Size of Crew
- Delays/Interruptions
- Elevations
- Site Congestion
- Stacking of Trades
- Multiple Floors
- Accessibility
- Pre-Fabrication

Chapter 6 gives more information on how to make adjustments to standard man-hours and how to calculate your own man-hours.

In **Figure 3.4**, a sample labor pricing sheet is given. Please note the following need to be added to the total labor costs:

- Medical insurance (health, dental, life, and disability)
- Tax-deferred pension or profit sharing plans
- Social security and medicare taxes (FICA)
- Federal Unemployment Tax (FUTA)
- State Unemployment Tax (SUTA)
- Worker's compensation insurance

PRICING OVERHEAD

Overhead is one of your costs, not profit. You must pay or "recover" your overhead each year by doing enough business to pay for it. You are not making any profit until all the overhead costs are recovered.

There are two types of overhead: jobsite overhead and office overhead. Each is calculated differently. Do not try to apply a flat rate covering both, as that is not accurate.

Jobsite Overhead

Jobsite overhead refers to costs directly related to your specific job. It is true that general contractors pay for many of these costs, such as toilet rental, material storage, etc. But, inevitably you will have to pay some jobsite overhead items from your own pocket. Normally the longer the job, the more jobsite overhead costs will be.

Figure 3.5 and **3.6** provide two separate lists of typical jobsite overhead cost items paid by HVAC contractors and general contractors. Because you may or may not have to pay for these costs, please verify with job conditions. A big chunk of jobsite overhead for large HVAC work is equipment rental.

Because jobsite overhead varies from job to job, you should figure a list of items instead of applying a percentage. **Figure 3.7** gives an example of calculating such a list.

PRICING OVERHEAD *(cont.)*

Office Overhead

Office overhead can not be directly tied to a specific job. You must pay for these costs to remain in business, whether you have any jobs or not. Examples of office overhead items include:

- Owner's Salary
- Salaries and Fringe Benefits for Office Personnel (e.g. estimators, draftsmen, bookkeepers, secretaries, etc.)
- Non-Job Vehicles, Fuels, and Insurance
- Office Rent, Utilities, Furniture, and Supplies
- Small Tools and Equipment
- Business License and Membership Dues
- Marketing and Advertising
- Loan Interest
- Legal and Auditing Expenses
- Taxes and Donations
- Bad Accounts

Office overhead varies from year to year. The best bet is to look at how much work you did last year and calculate the overhead you paid. Figure a percentage based on that, and then apply that percentage to the current estimate. Refer to **Figure 3.8** for an example of calculating office overhead costs.

ESTIMATING PROFIT

Profit is the money you want to make from the job, and it is normally estimated by applying a rate to the total cost. The rate could run 20% to 25% for small jobs and 10% to 15% on a large one.

In deciding the rate to be used, it is important to have a clear understanding of what and who you are up against. Evaluate how much risk you are taking and how much money you could make. Then decide what percentage gives you the greatest chance of you winning the job and making a profit. Be careful not to bid so low that you win the job but risk losing money. Your goal is to make a decent profit.

One of the best ways to determine the profit you can expect in today's competitive market is to look at the trend on your completed jobs. Keep a chart of completed jobs handy (as shown in Figure 3.9), and evaluate each job's actual performance against as-bid estimate. Did you realize the profit you desired very often? Then decide what sort of profit you can hope for on your current bid.

Profit is not a dirty word. Bid the project with a planned profit!

GETTING A CORRECT TOTAL

Now it is time to put together a correct total bid price. This is the scary "make-break" moment, as any cost items you missed will come out of your profit later.

A quick run-down of major cost areas:

- Material
- Labor
- Work by Subcontractors (see evaluation worksheet in **Figure 3.10**)
- Jobsite Overhead (including equipment, bond, insurance, permit, etc.)
- Office Overhead
- Owner's Allowance (check specs)
- Contingency (bad design documents, material/labor escalation, unforeseeable field conditions)
- Profit

Shown in **Figure 3.11** is a Summary Sheet for figuring total job costs, direct and indirect. Please keep in mind that the best way to avoid errors is to prepare as detailed an estimate as possible. Shortcuts frequently cause mistakes. Fast approaches such as mark-up tables (**Figure 3.12**) and combined pricing worksheets (**Figure 3.13**) should always be used with caution.

A final word: have someone else double check your finished estimate to reduce errors.

FIGURE 3.1 — MATERIAL PRICING WORKSHEET

Job Name:			
Ref No.:			
Estimator:			
Estimate Date:			
Sheet Numbers:			

Item	Quantity	Material Unit Price	List Price Subtotal
List Price Subtotal			
Discount			
Net Total			
Sales Tax			
Freight			
Total Material Costs			

The math in this sheet is as follows:

• List Price Subtotal = Quantity × Material Unit Price

• Net Total = List Price Subtotal − Discount

• Sales Tax = Net Total × Tax Rate

• Total Material Costs = Net Total + Sales Tax + Freight

FIGURE 3.2 — MATERIAL SHIPPING TERMS

F.O.B. Factory (Free On Board at the Factory): Title passes to you (the buyer) when the goods are delivered by the seller to the freight carrier. You pay the freight and are responsible for freight-damage claims.

F.O.B. Factory F.F.A. (Free On Board at the Factory, Full Freight Allowed): The title passes to you (the buyer) when the goods are delivered by the seller to the freight carrier. The seller pays the freight charges, but you are responsible for freight-damage claims.

F.O.B. Job Site (Free On Board at job site): The title passes to you (the buyer) when the goods are delivered to the job site (or shop). The seller pays the freight and is responsible for freight-damage claims.

F.A.S. Port (Free Alongside Ship at the nearest port): The title passes to you (the buyer) when goods are delivered to the ship dock or port terminal. The seller pays the freight and is responsible for freight-damage claims to the ship dock or port terminal only. You pay the freight and are responsible for freight-damage claims from the ship dock or port terminal to the designated delivery point.

Estimating Example:

If the shipping rate is $25.00 per cwt
(hundredweight),

Then a 6-ton air handling unit is
$6 \times 2000 = 12,000$ lbs or 120 cwt

Shipping cost is $120 \times \$25.00 = \$3,000$.

FIGURE 3.3 — MATERIAL DISCOUNT TABLE

Discount	Net	Discount	Net	Discount	Net
10-10	0.8100	30-10-10-5	0.5386	45-10	0.4950
				45-10-5	0.4703
20	0.8000	35	0.6500	45-10-10	0.4455
20-5	0.7600	35-5	0.6175	45-10-10-5	0.4232
20-5-5	0.7220	35-5-5	0.5866		
20-10	0.7200	35-5-10	0.5558	50	0.5000
20-10-10	0.6480	35-10	0.5850	50-5	0.4750
		35-10-5	0.5558	50-5-5	0.4513
25	0.7500	35-10-10	0.5265	50-5-10	0.4275
25-5	0.7125	35-10-10-5	0.5001	50-10	0.4500
25-5-5	0.6769			50-10-5	0.4275
25-5-10	0.6413	40	0.6000	50-10-10	0.4050
25-10	0.6750	40-5	0.5700	50-10-10-5	0.3847
25-10-5	0.6413	40-5-5	0.5415		
25-10-10	0.6075	40-5-10	0.5130	55	0.4500
25-10-10-5	0.5771	40-10	0.5400	55-5	0.4275
		40-10-5	0.5130	55-5-5	0.4061
30	0.7000	40-10-10	0.4860	55-5-10	0.3848
30-5	0.6650	40-10-10-5	0.4617	55-10	0.4050
30-5-5	0.6318			55-10-5	0.3848
30-5-10	0.5985	45	0.5500	55-10-10	0.3645
30-10	0.6300	45-5	0.5225	55-10-10-5	0.3462
30-10-5	0.5985	45-5-5	0.4964		
30-10-10	0.5670	45-5-10	0.4703	60	0.4000

FIGURE 3.3 — MATERIAL DISCOUNT TABLE *(cont.)*

Discount	Net	Discount	Net	Discount	Net
60-5	0.3800	65-10-10-5	0.2693	75-10	0.2250
60-5-5	0.3610			75-10-5	0.2138
60-5-10	0.3420	70	0.3000	75-10-10	0.2025
60-10	0.3600	70-5	0.2850	75-10-10-5	0.1923
60-10-5	0.3420	70-5-5	0.2708		
60-10-10	0.3240	70-5-10	0.2565	80	0.2000
60-10-10-5	0.3078	70-10	0.2700	80-5	0.1900
		70-10-5	0.2565	80-5-5	0.1805
65	0.3500	70-10-10	0.2430	80-5-10	0.1710
65-5	0.3325	70-10-10-5	0.2308	80-10	0.1800
65-5-5	0.3159			80-10-5	0.1710
65-5-10	0.2993	75	0.2500	80-10-10	0.1620
65-10	0.3150	75-5	0.2375	80-10-10-5	0.1539
65-10-5	0.2993	75-5-5	0.2256		
65-10-10	0.2835	75-5-10	0.2138	85	0.1500

Estimating Example:

If the list price for a certain valve is $4.50/EA and there are 20 EA required,

Then the material cost for this type of valves is $4.50 × 20 = $90

The supplier provides a discount of 25-10-10

From the table, the resulting rate is 0.6075

Then the net cost for valves is: $90 × 0.6075 = $54.68

FIGURE 3.4 — LABOR PRICING WORKSHEET

Job Name:

Ref No.:

Estimator:

Estimate Date:

Sheet Numbers:

Item	Quantity	Man-hours	Total Man-hours	Hourly Wage	Cost Subtotal
Labor Cost Subtotal					
Fringe Benefits					
Payroll Taxes					
Total Labor Costs					

The math in this sheet is as follows:
- Total Man-hours = Quantity × Man-hours
- Labor Cost Subtotal = Total Man-hours × Hourly Wage
- Total Labor Costs = Labor Cost Subtotal + Fringe Benefits + Payroll Taxes

FIGURE 3.5 — SALARY CONVERSION TABLE

Per Hour	Per Week	Per Month	Per Year
$6.00	$240	$1,039	$12,470
$7.00	$280	$1,212	$14,549
$8.00	$320	$1,386	$16,627
$9.00	$360	$1,559	$18,706
$10.00	$400	$1,732	$20,784
$11.00	$440	$1,905	$22,862
$12.00	$480	$2,078	$24,941
$13.00	$520	$2,252	$27,019
$14.00	$560	$2,425	$29,098
$15.00	$600	$2,598	$31,176
$16.00	$640	$2,771	$33,254
$17.00	$680	$2,944	$35,333
$18.00	$720	$3,118	$37,411
$19.00	$760	$3,291	$39,490
$20.00	$800	$3,464	$41,568
$21.00	$840	$3,637	$43,646
$22.00	$880	$3,810	$45,725
$23.00	$920	$3,984	$47,803
$24.00	$960	$4,157	$49,882
$25.00	$1,000	$4,330	$51,960
$26.00	$1,040	$4,503	$54,038
$27.00	$1,080	$4,676	$56,117
$28.00	$1,120	$4,850	$58,195
$29.00	$1,160	$5,023	$60,274

Note: This table is based on a 40-hr week, 4.33-week month of a 52-week year.

FIGURE 3.5 — SALARY CONVERSION TABLE *(cont.)*			
Per Hour	**Per Week**	**Per Month**	**Per Year**
$30.00	$1,200	$5,196	$62,352
$31.00	$1,240	$5,369	$64,430
$32.00	$1,280	$5,542	$66,509
$33.00	$1,320	$5,716	$68,587
$34.00	$1,360	$5,889	$70,666
$35.00	$1,400	$6,062	$72,744
$36.00	$1,440	$6,235	$74,822
$37.00	$1,480	$6,408	$76,901
$38.00	$1,520	$6,582	$78,979
$39.00	$1,560	$6,755	$81,058
$40.00	$1,600	$6,928	$83,136
$41.00	$1,640	$7,101	$85,214
$42.00	$1,680	$7,274	$87,293
$43.00	$1,720	$7,448	$89,371
$44.00	$1,760	$7,621	$91,450
$45.00	$1,800	$7,794	$93,528
$46.00	$1,840	$7,967	$95,606
$47.00	$1,880	$8,140	$97,685
$48.00	$1,920	$8,314	$99,763
$49.00	$1,960	$8,487	$101,842
$50.00	$2,000	$8,660	$103,920
$51.00	$2,040	$8,833	$105,998
$52.00	$2,080	$9,006	$108,077
$53.00	$2,120	$9,180	$110,155

Note: This table is based on a 40-hr week, 4.33-week month of a 52-week year.

FIGURE 3.5 — SALARY CONVERSION TABLE (cont.)

Per Hour	Per Week	Per Month	Per Year
$54.00	$2,160	$9,353	$112,234
$55.00	$2,200	$9,526	$114,312
$56.00	$2,240	$9,699	$116,390
$57.00	$2,280	$9,872	$118,469
$58.00	$2,320	$10,046	$120,547
$59.00	$2,360	$10,219	$122,626
$60.00	$2,400	$10,392	$124,704
$61.00	$2,440	$10,565	$126,782
$62.00	$2,480	$10,738	$128,861
$63.00	$2,520	$10,912	$130,939
$64.00	$2,560	$11,085	$133,018
$65.00	$2,600	$11,258	$135,096
$66.00	$2,640	$11,431	$137,174
$67.00	$2,680	$11,604	$139,253
$68.00	$2,720	$11,778	$141,331
$69.00	$2,760	$11,951	$143,410
$70.00	$2,800	$12,124	$145,488
$71.00	$2,840	$12,297	$147,566
$72.00	$2,880	$12,470	$149,645
$73.00	$2,920	$12,644	$151,723
$74.00	$2,960	$12,817	$153,802
$75.00	$3,000	$12,990	$155,880
$76.00	$3,040	$13,163	$157,958
$77.00	$3,080	$13,336	$160,037

Note: This table is based on a 40-hr week, 4.33-week month of a 52-week year.

FIGURE 3.5 — SALARY CONVERSION TABLE *(cont.)*

Per Hour	Per Week	Per Month	Per Year
$78.00	$3,120	$13,510	$162,115
$79.00	$3,160	$13,683	$164,194
$80.00	$3,200	$13,856	$166,272
$81.00	$3,240	$14,029	$168,350
$82.00	$3,280	$14,202	$170,429
$83.00	$3,320	$14,376	$172,507
$84.00	$3,360	$14,549	$174,586
$85.00	$3,400	$14,722	$176,664
$86.00	$3,440	$14,895	$178,742
$87.00	$3,480	$15,068	$180,821
$88.00	$3,520	$15,242	$182,899
$89.00	$3,560	$15,415	$184,978
$90.00	$3,600	$15,588	$187,056
$91.00	$3,640	$15,761	$189,134
$92.00	$3,680	$15,934	$191,213
$93.00	$3,720	$16,108	$193,291
$94.00	$3,760	$16,281	$195,370
$95.00	$3,800	$16,454	$197,448
$96.00	$3,840	$16,627	$199,526
$97.00	$3,880	$16,800	$201,605
$98.00	$3,920	$16,974	$203,683
$99.00	$3,960	$17,147	$205,762
$100.00	$4,000	$17,320	$207,840

Note: This table is based on a 40-hr week, 4.33-week month of a 52-week year.

FIGURE 3.6 — TYPICAL JOBSITE OVERHEAD ITEMS PAID BY HVAC CONTRACTOR

- Bond for work under this trade
- Permit and license for HVAC work
- Liability Insurance for work under this trade
- Job Mobilization
- Tool Sheds
- Supervision
- Travel Expenses
- Job Vehicles and Fuels
- Temporary Water
- Weather Protection
- Jobsite Signs
- Trade Clean-up
- Punch-list Items
- Shop Drawings
- Product Sample Submittals
- Surveys for HVAC work
- As-builts
- Callback during warranty period

FIGURE 3.7 — TYPICAL JOBSITE OVERHEAD ITEMS PAID BY GENERAL CONTRACTOR

- Building Permit
- Job Trailer
- Temporary Power
- Jobsite Monthly Utilities (heat, electricity, gas, water, telephone, etc.)
- Trades Parking
- Temporary Toilets
- Material Storage
- Temporary Partitions and Enclosure
- Fencing, Gates, and Barricade
- Drinking Water
- Safety Equipment and First Aid
- Drawing Reproduction
- Surveying and Layout
- Miscellaneous Cutting and Patching
- General Jobsite Daily Clean-up
- Final Clean-up on Completion
- Monthly Progress Photos

FIGURE 3.8 — ESTIMATING JOBSITE OVERHEAD

This is the sample jobsite overhead estimate for a small HVAC job:

Item	QTY	Unit	Rate	Subtotal
Bond	1	l/s	$1,000	$1,000
Liability insurance	1	l/s	$1,200	$1,200
Permit	1	l/s	$500	$500
Mobilization	1	l/s	$400	$400
Superintendent	8	wk	$1,500	$12,000
Foreman	8	wk	$800	$6,400
Hotel Expenses	8	wk	$500	$4,000
Project sign	1	l/s	$500	$500
Telephone bills	2	mo.	$150	$300
Temporary water	1	l/s	$600	$600
Vehicles & fuels	2	each	$1,200	$2,400
Hoisting	1	wk	$900	$900
Small tools	1	l/s	$500	$500
Submittals	1	l/s	$200	$200
Daily clean-up	40	day	$40	$1,600
Punch-list items	1	l/s	$1,000	$1,000
As-built drawings	1	l/s	$300	$300
Total Jobsite Overhead				**$33,800**

FIGURE 3.9 — ESTIMATING OFFICE OVERHEAD

Formula:

Rate = Office Overhead last year/Construction Volume last year

Total Direct Costs for current bid = Material + Labor + Jobsite Overhead

Office Overhead for current bid = Rate × Total Direct Costs for current bid

Example:

Office Overhead last year: $30,000

Construction Volume last year: $500,000

Office Overhead rate: $30,000/$500,000 = 6%.

Total Direct Costs for your current bid: $190,000

Office Overhead for current bid: $190,000 × 6% = $11,400.

FIGURE 3.10 – DETERMINING PROFIT RATE

Job Name:

Ref No.:

Estimator:

Estimate Date:

Sheet Numbers:

Job Name	Contract Amount	Job Superintendent	As-Bid Profit Rate	Actual Profit Rate	Rate Evaluation
Today's Rate					

FIGURE 3.11 — SUBCONTRACTOR QUOTE EVALUATION WORKSHEET

Job Name:	
Ref No.:	
Estimator:	
Estimate Date:	
Sheet Numbers:	
Items Quoted	
Subcontractor Name	
Contact Name/Phone No.	
Work Included	**Amount**
Total Base Bid Amount	
Delivery Time	
Does the quote include the following	**Yes/No/Amount**
Sales Tax	
Delivery to Jobsite	
Complete Installation	
Per Plans and Specs	
Excluded Items	**Adjusted Amount**
Total Adjustment to Base Bid	
Addenda Acknowledgement	
Alternate No	**Amount**

FIGURE 3.12 — PRICING SUMMARY WORKSHEET

Item	Rate	Amount	Subtotal
Material Subtotal			
Labor Subtotal			
Subcontractor			
Jobsite Overhead			
Total Direct Costs			
Office Overhead			
Owner's Allowance			
Contingency			
Total Price Before Profit			
Profit			
Total Price			

The math in this sheet is as follows:

- Total Direct Costs = Material Subtotal + Labor Subtotal + Subcontractor + Jobsite Overhead

- Office Overhead = Total Direct Costs × Office Overhead Percentage

- Total Price Before Profit = Total Direct Costs + Office Overhead + Owner's Allowance + Contingency

- Profit = Total Price Before Profit × Profit Rate

- Total Price = Total Price Before Profit + Profit

FIGURE 3.13 — MARKUP TABLE

		\multicolumn{11}{c}{Office Overhead Rate}									
		15%	16%	17%	18%	19%	20%	21%	22%	23%	24%	25%
	5%	0.80	0.79	0.78	0.77	0.76	0.75	0.74	0.73	0.72	0.71	0.70
	6%	0.79	0.78	0.77	0.76	0.75	0.74	0.73	0.72	0.71	0.70	0.69
	7%	0.78	0.77	0.76	0.75	0.74	0.73	0.72	0.71	0.70	0.69	0.68
	8%	0.77	0.76	0.75	0.74	0.73	0.72	0.71	0.70	0.69	0.68	0.67
	9%	0.76	0.75	0.74	0.73	0.72	0.71	0.70	0.69	0.68	0.67	0.66
	10%	0.75	0.74	0.73	0.72	0.71	0.70	0.69	0.68	0.67	0.66	0.65
	11%	0.74	0.73	0.72	0.71	0.70	0.69	0.68	0.67	0.66	0.65	0.64
Profit Rate	**12%**	0.73	0.72	0.71	0.70	0.69	0.68	0.67	0.66	0.65	0.64	0.63
	13%	0.72	0.71	0.70	0.69	0.68	0.67	0.66	0.65	0.64	0.63	0.62
	14%	0.71	0.70	0.69	0.68	0.67	0.66	0.65	0.64	0.63	0.62	0.61
	15%	0.70	0.69	0.68	0.67	0.66	0.65	0.64	0.63	0.62	0.61	0.60
	20%	0.65	0.64	0.63	0.62	0.61	0.60	0.59	0.58	0.57	0.56	0.55
	25%	0.60	0.59	0.58	0.57	0.56	0.55	0.54	0.53	0.52	0.51	0.50
	30%	0.55	0.54	0.53	0.52	0.51	0.50	0.49	0.48	0.47	0.46	0.45
	35%	0.50	0.49	0.48	0.47	0.46	0.45	0.44	0.43	0.42	0.41	0.40
	40%	0.45	0.44	0.43	0.42	0.41	0.40	0.39	0.38	0.37	0.36	0.35
	45%	0.40	0.39	0.38	0.37	0.36	0.35	0.34	0.33	0.32	0.31	0.30
	50%	0.35	0.34	0.33	0.32	0.31	0.30	0.29	0.28	0.27	0.26	0.25

Estimating Example:

If the direct costs (labor, material, and job overhead) add up to $25,000

You want to apply 18% of that as office overhead

You want to apply 10% of that as profit

From the table, the "mark-up" rate is 0.72

The result can be found by: $25,000/0.72 = $34, 722

Adding Owner's Allowance and Contingency will give you a final bid price.

Note: Please use this speedy estimating method with caution.

FIGURE 3.14 – COMBINED PRICING WORKSHEET

Item/Description	Quantity	Unit	Material Unit Price	Material Subtotal	Man-Hours	Total Man-Hours	Labor Unit Price	Labor Subtotal
Material Discount								
Sales Tax								
Freight								
Fringe Benefits								
Payroll Taxes								
Subcontractor								
Jobsite Overhead								
Total Direct Costs								
Office Overhead								
Owner's Allowance								
Contingency								
Total Price Before Profit								
Profit								
Total Price								

Note: Please use this speedy estimating form with caution. This worksheet is more helpful when you are bidding a small job and do not have a lot of items to estimate. For best estimating results, please follow the step-by-step approach as described in this chapter.

CHAPTER 4
Bidding

Bid day can be quite hectic, both for those who prepare the price proposals and for those who receive them. No matter how prepared you may be, there could be surprises or problems. Choose to be organized as organization helps reduce errors as well as stress.

WRITING A PROPOSAL

With a total price calculated, you are now ready to submit a proposal to your customer. An example proposal form is provided in **Figure 4.1**. In writing a proposal, choose your words carefully (e.g. stating how long the price is good and how you expect to be paid when the work starts).

Include at least the following information on your proposal:

- Proposal number, date, and version of revision
- Complete job name and address
- Name of owner, architect/engineer, or general contractor
- A complete list of drawings/specifications/addenda with issue or revision date
- Base bid price as well as the numbers for alternates
- A list of inclusions, exclusions, clarifications, and assumptions
- A standard contract from your company
- Contact name and phone number

PROPOSAL INCLUSION AND EXCLUSIONS

Some HVAC contractors would rather keep their estimating information as secret as possible. Thus their proposals are no more than a lump sum price with a short explanation of "Per Plans and Specs." This approach will not help you stand out from the competition bidding the job. There are two main reasons:

1. General contractors will always need to call you with a number of questions demanding explanations before they decide who to use. "Per Plans and Specs" means more questions. Some general contractors will not bother calling if another proposal has a close number and has everything outlined in detail. So, unless you have good relationship with the general contractor, no one will have reason to believe that you have everything covered.

2. Frequently plans and specs are not perfect. When the job starts, the general contractor may approach you as the low bidder. Then he may discover that there are a lot of scope issues he has to go through with you. That is not a pleasant situation for anyone. So, if you do not want to enter a contract only based on "per plans and specs," then do not do a price proposal that way in the first place.

It is best to be clear about what your prices are based on. **Figures 4.2** thru **Figure 4.9** give a standard list of inclusions and exclusions for your reference.

VALUE ENGINEERING IN BIDDING

"Value engineering," cost saving through changing the design, is often regarded as a way to help owners save money. If properly used in your price proposal, value engineering ideas can make people think of you as "team player," thus increasing your potential for getting a job. For example, if you believe a smaller boiler than the one currently designed could be used to sufficiently meet the demand of building, it could result in some cost savings.

Take the following advice:

1. Always separate value engineering prices from your base bid. Even if you think you can design buildings better than the engineer, do not try to reflect that in the base bid, which should be based on plans and specs as well as the codes.

2. Do not waste your time offering to use cheaper or lower-quality material. This is not value engineering, but an insult to your customer. The money saving should be done without sacrificing the system quality or capacity (or even as an improvement with less money).

3. Include your overhead when proposing value engineering numbers. For example, if a suggested design change can save the owner $10,000 in total, you should only propose a smaller amount of savings (say $7,000). There may be additional work for you to do because you are dealing with an uncertain thing.

LIFE CYCLE COST ANALYSIS

Life cycle costing is a technique considering the total costs of building systems, both at the present time and in the future. It is more commonly used for "design-build" or negotiated jobs, where contractors are to work on both design and costing. In these situations, you try to convince the customer that the proposed design will save them money over time, although it may require a more expensive initial investment.

Imagine you propose a new geo-thermal system which costs $200,000 to construct. But if a conventional heating system were used, it would cost only $125,000. The owner may wonder why you would recommend a more expensive design option ($75,000).

In response, you admit that it is true that the proposed design is more expensive in initial investment, but the new system will save money in the long run: lower utility costs each year, easier to maintain, longer lasting, cheaper to operate, more environment-friendly, eligible for tax credits, etc. You can also point out that the current construction cost difference of $75,000 will be made even within a few years. After that, the system generates significant net cost savings annually.

In using a life cycle costing method, it is important to show the customers that the benefits of your proposed system outweigh the excessive costs in the long run. That way, they can be sure that your proposal was taking their needs into consideration, not trying to intentionally make it more expensive just for your own profit.

COST BREAKOUT REQUIREMENTS

Owners normally put their cost breakdown requirements into the specs. For example, if there are two buildings, they may want to know the cost for each. HVAC contractors bidding a job should verify such requirements with general contractors or owners and prepare proposals accordingly.

However, sometimes contactors are not informed or completely forget about the need for cost breakout until they are done with all estimating and pricing. The client may ask for the total prices to be broken out into several smaller pieces. There are a couple of shortcuts which can be used to avoid redoing the estimate completely.

A common starting point is breaking out by the gross building area (i.e. square feet). Suppose there are quite a few buildings in one job. It normally makes sense that the larger ones should have a larger share of the total costs, if the buildings are quite similar.

Another method is breaking out by the actual functional components. For residential or commercial HVAC work, estimators frequently use the number of major equipment like roof top units.

Example calculations are shown in **Figure 4.10** for these two methods.

SUBMITTING THE PROPOSAL

Figure 4.11 provides a list of issues to check before you send out a proposal. Your final bid can be phoned in, sent by fax, or e-mailed. Sending it by regular mail is normally too slow and adds the risk of your proposal getting lost.

Before the bid day, it is a good idea to send out a "scope quote" with everything completed except the price. This is to assure the general contractor that you are still bidding and will send a formal proposal the next day. It also gives him some clues about what you will be bidding on. For example, if you don't want to include test and balance, then he will know that in advance and make allowances to cover it.

Try to turn in your proposal before the deadline required. If you cannot make a bid or need more time, always call the general contractor or the owner as soon as you can to discuss the situation. Do not wait until the last minute. After you send your proposal, give the receiving party a brief phone call to confirm they did receive it, and get the name of the contact person who will evaluate HVAC quotes.

The general contractor might contact you on the bid day, either because there are mistakes in your proposal (e.g. math errors), or to clarify what you actually include or exclude (perhaps also wanting you to do something about that). Make yourself available to answer questions; otherwise they might simply use another guy's price.

COPING WITH BID SHOPPING

It is true that in today's construction industry, there are some owners and general contractors who do not run their business morally. They can use your already low quote to solicit lower bids from your competition, or pressure you into cutting your price by threatening to award the work to others. This unethical practice is called "bid shopping," and actually can occur either during or after the bid (when a general contractor has the job).

Some HVAC contractors become part of "bid shopping" partly because they want the job too much. They inquire about details of their competition's quote and offer to beat that price. They may also guarantee to provide a figure which is a certain percentage below the current lowest bid. However, a job that is profitable for one contractor may be unprofitable for another. Essentially, it is a gamble to guarantee a price cut.

Generally, the truly successful contractor (general and his subs) do not engage in bid shopping. The real solution is to know who you are bidding with. If a general contractor or owner has a bad reputation of bid shopping, perhaps it's wise not to bid jobs for him.

FIGURE 4.1 — BID PROPOSAL FORM

Proposal No.: _____   Revision No. _____

Date: _____

From: _____

To: _____

Project Name & Address: _____

_____

We hereby propose to furnish materials and perform the labor necessary for the construction of HVAC systems for the above project. The work is to be performed in accordance with drawings _____ and specifications _____ we received to date. Receipt of Addendum # _____ is acknowledged.

Base Bid: _____   Dollars ($ _____)

Alternate: _____   Dollars ($ _____)

Our bid includes: _____

Our bid excludes: _____

This proposal is valid for ____ days from the date of the submission.

Respectfully submitted by:

Signature: _____

Title:_____

FIGURE 4.2 — GENERAL SCOPE INCLUSIONS

- Items required by codes regardless of where they appear on the bid documents

- Contingency funds specifically called for in the specifications

- Cash allowances defined in the specifications

- All service charges as levied by the town, city, county, or utility company

- Sales tax

- Permits, licenses, and liability insurances

- Trade clean-up

- Keeping job free of personal garbage

- Removal and disposal of surplus materials

- Protection of other trades' work from damage

- Rigging, hoisting, and placing of equipment or materials supplied by this trade

- Vibration isolation for equipment supplied by this trade

- Scaffolding required by the work of this trade

- Removal or relocation of existing mechanical components

- Pre-marking the locations of penetrations through other trades' material

FIGURE 4.2 — GENERAL SCOPE INCLUSIONS (cont.)

- Cutting penetrations through walls, floors, and ceilings for 6" and smaller pipes and ducts

- Cutting recesses in walls which do not exceed 32" × 32"

- Installation of sleeves for penetrations through other trades' material.

- Fire-stopping materials at drilled holes, sleeves, and other openings to maintain integrity of fire separations

- Patching and making good materials related to openings under this trade

- Shop drawings, samples, mockups, as-built drawings, maintenance data, and operating instructions to owners

- Anchor bolts, supports, and frames for mechanical equipment

- Supply of access panels or clean out covers for pipe, ducting, and equipment

- Identification, adhesive markers, valve tags, labels, and framed directories

- Supply and installation of prefabricated curbs for roof top units

- Testing and inspections

- Guarantees and warranties

FIGURE 4.3 — HEATING/AIR CONDITIONING SCOPE INCLUSIONS

- Gas fired, oil fired, and electric boilers
- Factory assembled centrifugal packaged reciprocal and screw type water chillers complete with starters, factory start up, and service
- Cooling towers and evaporative water coolers
- Pumps
- Furnaces including gas and oil fired
- Gas and oil fired duct heaters and unit heaters
- Roof top units, air handling units, and components
- Reheat, preheat, and air tempering coils (water, steam, or glycol)
- Wall fin elements, induction units, fan coil units, force flo, convectors, unit heaters, and the associated metal enclosures
- Humidifiers: water, steam, or electric
- Self-contained packaged terminal heating or air conditioning units or both
- Absorption chillers
- Energy storage tanks
- Packaged heat pumps
- Self-contained control devices and gauges
- Incinerators including pathology incinerators complete with gas or oil burners, draft inducers, breeching, and stacks
- All wet and dry type solar heating and air conditioning systems complete with associated accessories
- Fuel tanks for power plant equipment complete with dispensing units and piping

FIGURE 4.4 — REFRIGERATION SCOPE INCLUSIONS

1. Job site assembled split systems using all refrigerants (except water) for air conditioning for human comfort. Include the complete refrigeration circuit and components as follows:

- Refrigeration compressors complete with motors, starters, and operating controls
- Condensers (Evaporative, water-cooled, or air-cooled)
- Evaporators: Primary liquid coolers (split primary refrigerant chillers)
- Primary refrigerant pressure vessels and primary refrigerant piping
- Primary refrigerant safety controls and wiring for these controls when they form an integral part of the factory assembled unit (known as a packaged unit)
- Primary refrigerant operated water regulating valves
- Labor and equipment to rig and place primary refrigeration
- Piping vibration elimination equipment
- Specified Original Equipment Manufacturer's (OEM) bases and supports
- Special rack or steel bases and supports for equipment required by this trade
- Charging and testing of all primary refrigerant circuits

FIGURE 4.4 — REFRIGERATION SCOPE INCLUSIONS *(cont.)*

• Heat reclaim coil when associated with primary refrigerants for air conditioning for human comfort

2. Packaged air conditioning units and refrigerated air dryers (other than control system air dryers):
- Packaged air conditioning units with primary refrigeration circuits
- Packaged air conditioning units with special accessories
- Packaged unit temperature and humidity controls and control wiring
- Fan coil units when part of a split air conditioning system containing DX coils, fans, filters, and motors interconnecting refrigeration piping
- Split system heat pumps
- Roof top and air handling units and components as related to refrigeration

3. Evaporative condensers containing primary refrigerant coils:
- Fans, motors, drives, and accessories
- Supports and installation of vibration isolation units
- Evaporative controls
- Water treatment equipment for evaporative condensers installed directly in unit and not fitted in the piping

FIGURE 4.4 — REFRIGERATION
SCOPE INCLUSIONS *(cont.)*

4. Product Refrigeration

The supply and installation of job site assembled systems (using primary refrigerants for product refrigeration, special cold rooms, water coolers, ice makers, and process mechanical cooling) will include the complete refrigeration circuit and all components listed below:

- Refrigeration compressors complete with motors and operating controls
- Condensers (evaporative, water-cooled, or air-cooled), complete with motors
- Evaporators (force flow, gravity, pipe coil, or plate type), complete with motors
- Primary refrigerant pressure vessels and refrigerant piping
- Primary refrigerant safety controls, all devices other than automatic stop/start control contact
- Primary refrigerant
- Specified OEM bases and supports
- Special racks or steel bases and supports for equipment required by this trade
- Charging and testing of primary refrigerant circuits
- Primary refrigerant pipe
- Low temperature defrost systems and controls
- Low temperature condensate drain line heaters
- Direct expansion coils mounted in ductwork
- Secondary (water) flow control switches for air conditioning equipment

FIGURE 4.5 — VENTILATION SCOPE INCLUSIONS

- Sheet metal and non-metallic air handling ductwork of all types including hangers and supports

- Back draft dampers, manual dampers, fire dampers, ceiling fire flaps, motorized smoke dampers, registers, grilles, diffusers, screens, and air troffer boots.

- Breechings, vents, chimneys, thimbles, and steel chimneys - include draft regulators and clean outs

- Plenums and casing

- Carbon monoxide systems

- Air curtains

- Exhaust hoods and canopies including kitchen type hoods when there is no food service equipment trade on the project

- Brand named fume hoods and flow hoods only when shown on the mechanical drawings and in the mechanical specification

- Dust collectors

- Paint spray booths

- Access doors pertaining to sheet metal work

- Internal duct insulation

- Freestanding or "dry" fans, including return air fans and exhaust fans

FIGURE 4.5 — VENTILATION
SCOPE INCLUSIONS *(cont.)*

- Roof ventilators, gravity exhausters, and intake hoods (including metal curbs, duct connected, dampered, or both)

- Induced draft fan

- Ceiling prop fans including guards and switches

- Supports for equipment required by this trade

- Vibration isolation materials for equipment required by this trade

- Air filters and filter gauges

- Duct silencers and sound attenuation equipment

- Terminal re-heat boxes, variable volume units, dual duct boxes, and pressure reducing air valves

- Solar energized forced air system

- Electric unit heaters, duct connected

- Electric in line duct heaters, not independently supported but duct connected

- Packaged room cabinet type electric heating unit requiring air intake and distribution duct to be connected

- Completely assembled electric furnaces or disassembled multi-sectional electric heating furnaces, whether or not duct connected, consisting of fan sections and/or mixing box section and/or electrical element section and/or filter section

FIGURE 4.5 — VENTILATION
SCOPE INCLUSIONS *(cont.)*

- Burglar bars totally enclosed by ductwork

- Packaged controls for variable air volume (VAV) boxes

- Motorized fire and fire/smoke combination dampers including actuator

- Variable speed drives when specified to be factory supplied and wired with packaged equipment

- Control dampers for motorized actuation

- Installation of air flow measuring stations

- Wood burning furnaces

- Ducted dehumidifiers

- Air blenders not included in air handling units

- Drain/drip pans, troughs, and liners where shown in ductwork, under fans, relief air openings, and mechanical piping

- Waste gas scavenger exhaust air systems

- Louvers and pre-manufactured louvered pent-houses

- Roof top and air handling units and components

- Vacuum cleaning of existing ducts

- Vibration isolation for equipment supplied by other mechanical trades

- Direct expansion coils mounted in ductwork

FIGURE 4.6 — AUTOMATIC TEMPERATURE CONTROL SCOPE INCLUSIONS

Environmental Control
- Heating controls
- Ventilation controls complete with actuators for fan volume control
- Cooling controls
- Humidification controls
- Air conditioning controls
- Refrigeration controls, (devices used for automatic system stop/start control contact only), ambient lockout control device unless supplied complete with packaged system
- Boiler controls (not specified as part of pre-wired package boiler)
- Motorized dampers
- Remote reading gauges, transmitters, and recorders for automatic recorders
- Control compressor, air driers, PRV's filters, and isolators

Building Centralization Control
- Centralized temperature, humidity, pressure, and flow indication and recorders
- Centralizations of air conditioning, heating, ventilation, humidification, and refrigeration equipment operation

Building Life Safety Control and Equipment Safety Control
- Motorized smoke dampers
- Motorized fire dampers
- Firestats
- Freeze protection thermostats
- Carbon monoxide detection systems (i.e. carbon dioxide, NOX, propane, diesel)
- Shutdown safety devices forming part of the automatic temperature control

FIGURE 4.7 — INSULATION SCOPE INCLUSIONS

❏ Insulation for:
- Heating piping, vessels, and equipment
- Cooling piping, vessels, and equipment
- Refrigeration piping, vessels, and equipment
- Boilers, hot water heaters, water storage and vessels, convectors, breeching, stacks, and furnace flues
- Chillers and evaporators
- Emergency generator exhausts
- External application of materials for fire rating of plenums, grease ducts, etc., including rating of support systems
- External application to the ductwork, fittings, and equipment

❏ Insulation finishes and weatherproofing

❏ Installation of pipe covering protection shields

❏ Underground insulation conduits

❏ Insulation above radiant ceiling panels (active or non active)

FIGURE 4.8 — TESTING AND BALANCING SCOPE INCLUSIONS

❑ Air and fluid flow measurements, testing and distribution (balancing) relevant to heating, ventilation, and air conditioning (HVAC) systems

❑ Air flow duct leakage tests

❑ Sound level and equipment vibration testing readings

❑ Testing of fire dampers and fire smoke dampers

- Verify the unit is fully accessible
- Unit has been successfully tested and reset
- Tag attached with name of tester and date tested
- Report including location, access, size, link rating, and date unit tested successfully

❑ Indoor air quality test

❑ Installation only of required drive and belt changes

FIGURE 4.9 — GENERAL SCOPE EXCLUSIONS

- Any and all portion of the building permit
- Major material price increases
- Temporary heat, light, power, water, and sanitation
- Project trailer or office facilities
- Project insurance
- Jobsite security
- Fencing and safety enclosures
- Exterior hoarding and heating
- General survey and layout
- Reasonable access to site, including snow clearing
- Distribution and review of samples, shop drawings, and other submissions
- Protection of building finishes (i.e. floors, wall surfaces, ceilings, etc.)
- Independent inspections
- Certified test and balance
- Material storage costs
- Hauling of site trash
- Final clean-up
- Backings or supports for equipment
- Asbestos removal

FIGURE 4.9 — GENERAL SCOPE EXCLUSIONS *(cont.)*

- Electrical work including wiring, conduit, and motor starters
- Plumbing work unless noted otherwise
- Fire protection and irrigation sprinklers
- Concrete work, including formwork and reinforcement
- Painting, priming, and surface preparation
- Plaster, caulking, and grouting of all types
- Structural cutting, patching, or repairing
- Architectural louvers
- Insulation not applied to HVAC systems
- Garbage and linen chutes
- Catwalks, ladders, grating, and support unless required to be done by this trade
- Supply or installation of FFE (Furniture, Fixture, and Equipment)
- All residential appliances
- Prefabricated coolers and freezers
- All modular unit components
- Fire dampers not shown on plans

FIGURE 4.10 – COST BREAKOUT CALCULATION

Note: All dollar amounts in this example are fictional and only for illustrative purposes.

1. Based on Building Area
Estimating Example:

You total bid price is $50,000.

Building A has 1,500 SF and Building B has 2,500 SF

Then the total building area is: 1,500 + 2,500 = 4,000 SF

Percentage for Building A of the total area: $1,500/4,000 \times 100\% = 37.5\%$

Percentage for Building B of the total area: $2,500/4,000 \times 100\% = 62.5\%$

Therefore

Cost for Building A: 37.5% × $50,000 = $18,750

Cost for Building B: 62.5% × $50,000 = $31,250

2. Based on Functional Components
Estimating Example:

You total bid price is $60,000.

Building A has 2 roof top units 3 tons each, and Building B has 3 units 4 tons each

Then the total roof top units tonnages are: $2 \times 3 + 3 \times 4 = 18$ tons

Percentage for Building A of the total roof top units: $6/18 \times 100\% = 33.33\%$

Percentage for Building B of the total roof top units: $12/18 \times 100\% = 66.67\%$

Therefore

Cost for Building A: 33.33% × $60,000 = $20,000

Cost for Building B: 66.67% × $60,000 = $40,000

FIGURE 4.11 – QUESTION LIST BEFORE SENDING A BID

1. Have you double-checked the time and date for the bid to be submitted?
2. Does your bid include everything required in the scope?
3. Is your bid in accordance with plans and specs?
4. Is your bid in accordance with mechanical codes?
5. Does your bid exclude everything that should not be included?
6. Are you aware of project schedule requirements? Can you finish it on time?
7. If you worked the bid from an old estimate, have you made applicable changes?
8. Did you check your math from beginning to end?
9. Have your questions been answered by an architect, engineer, or general contractor?
10. Have you received and reviewed all addenda transmitted by the general contractor?
11. Did you breakout your quote in the format as required by the bid?
12. Are there any gaps in the quotes from your subs that you need to fill on your own?
13. Did you include bond, insurance, and permit?
14. Did you include required owner's allowance?
15. Did you include some contingencies for design issues, price escalation, and unforeseeable site conditions?
16. Did you fill out all the blanks including alternates as well as base price?
17. Is the bid proposal signed by the company owner?
18. Do you want to add a business letter or standard contract?
19. Can the general contractor reach you easily if he has any questions regarding the bid?
20. Did someone else double check your estimate and proposal?

CHAPTER 5
Post-Bid

Estimating does not end with faxing a copy of a proposal to the customer. A lot of things could happen after you submit a bid. Do not sit and wait for the phone to ring. Instead, take a proactive approach and you may land a great job!

POST-BID REVIEW

Immediately after the bid is over, sit down to review the bid you submitted. You may find a few things you did not do so well. **Figure 5.1** and **5.2** list some common estimating mistakes and ways to reduce them. Record-keeping is important, as you never know when a customer is going to call back with questions. Organize all the paperwork in a post-bid folder for future reference, including:

- Drawings, Specs, Addendums
- Bid Proposal
- Estimate Worksheets
- Quotes from Suppliers and Subs
- Bid Notes
- Site Visit Photos

Some analytical work with the price proposal will give you an idea about cost per square foot, etc. Use the worksheet in **Figure 5.3** to do this.

PROPOSAL FOLLOW-UP

Call the general contractor approximately two weeks after the bid and ask the following questions:

- Have you been awarded the job from the owner?
- Did you receive all the information you requested?
- Do you have any questions on what you received?
- Is there anything else you need at this point?
- When would be a good time to check back with you?
- Who is the project manager, if such person has been assigned to the job?
- Who is the job superintendent?
- Who makes the decision regarding subcontract award?
- What are the important selection criteria?
- How important is price as a selection factor?
- Could I schedule a meeting with you to discuss the proposal?

Emails can also be used as an alternative to phone calls in inquiring the status. Some people are more willing to respond to emails than phone calls, as emails do not interfere with what they are doing.

Although it is difficult to get a face-to-face meeting, you cannot do everything over the phone. Do whatever it takes to get a face to face meeting, including just showing up at their office and waiting in the lobby until they meet you. Prepare yourself well to make a good presentation.

IMPROVING BID-HIT RATIO

Studies show that HVAC contractors normally bid six or seven jobs before they get one. If you get every job you bid, you are bidding too low. On the other hand, if only one bid turns out to be successful for every 20 or 30 jobs you bid, you are spending too much time and effort estimating and not enough time working and earning a profit. Many HVAC contractors keep bidding to the same customers over and over, using the same strategy every time. At some point, they must realize that this does not work very well.

Developing a bid history and track it monthly, quarterly, and yearly may help. Track all types of projects you bid on and each customer you bid to: large versus small, hard bid versus negotiated, plans-specs versus design-build, new construction versus renovation, local versus out of town, commercial versus residential versus industrial, public versus private, etc.

Figure 5.4 offers a preliminary worksheet for you to customize and use. As you study these jobs, you will find certain customers give you more work than others. You will also discover you do better with certain kinds of jobs. This simple tracking method will help you to focus on the real customers who can give you the best jobs for you. For some jobs, you will be surprised to find out price is not the most important factor.

ESTIMATING FOR CONTRACT

If the general contractor is talking to you regarding your proposal, that indicates that at least your price is competitive, and you may be the low bidder. But, before offering or accepting a subcontract, a revised estimate needs to be done. The final HVAC contract is rarely 100% based on initial bid proposal, considering there are usually numerous post-bid changes.

Use the guidelines in **Figure 5.5** for preparing a revised post-bid estimate.

Once you finally are awarded the opportunity of a contract, be aware of what you sign. If you have your own "standard" contract, see if the general contractor or the owner is willing to accept it as an alternative to their contract. If this is not possible, read the contract they insist on. Read everything including the fine print. Do not assume that this contract is the same as the one in the original bid package. Verify it word for word. If something seems unfair to you, offer to cross it out. If they claim that the offensive portion is not relevant, you should reply that it is better then to eliminate it. If you want something included that is not there, write it on all copies and have them initial the addition.

Figure 5.6 gives major elements in a HVAC sub-contract, while in **Figure 5.7** there is an example for the HVAC scope of work.

In addition, you need to pay special attention to the following contract items:

- Payment Clauses
- Penalties and Indemnity Clauses
- Notice Provisions
- Change Order Procedures
- Dispute Resolution
- Breach of Contract
- Cancellation of Contract
- Mechanics Lien
- Payment and Performance Bonds
- Time Limit of Legal Rights

Figure 5.8 lists some negative clauses you should always avoid. In any event, do not start the work until you have your copy of the amended, countersigned, and initialed contract in your possession.

TURN-OVER MEETING

If you are lucky enough to get a seemingly profitable job, then the next thing you do is hold a "turn-over" meeting with your field personnel who will run the daily jobsite operation.

Why have "turn-over" meetings when everyone can just look at a copy of the job estimate for answers? This is necessary because by communicating with the estimator, project managers and superintendents are much better informed about the project. Then they are able to start the submittal process and place vendor and subcontractor purchase orders accurately, providing all the necessary materials and tools for a quick start.

To have a successful meeting, determine who should attend: estimator, project manager, superintendents, jobsite trade foreman, and job accountant. External personnel are normally not permitted, as "turn-over" meetings always involve sensitive information such as contract amount, low vendors and subcontractors, etc.

A "turn-over" meeting should be more than just handing over a package of plans and specs. Instead, follow a meeting agenda like the one in **Figure 5.9**.

FIGURE 5.1 — COMMON BIDDING MISTAKES

a. Not including required items: scope omission is perhaps the most serious mistake.

b. Simple math errors: you may have incorrectly added or subtracted numbers, or used wrong formula or conversion factors.

c. Measurement errors: you could have used the wrong scale for reduced-size drawings. For example, when drawings are half-size and you used the scale as shown, then your area was reduced by 75%, not the 50% reduction you may have thought.

d. Incorrect material prices: you fail to get material price updates from the suppliers and the unit prices you used are too old.

e. Insufficient labor coverage: too optimistic about certain items that will take a longer time to install; or crews won't be available when the job starts.

f. Underestimated job duration: you do not have enough money to cover jobsite overhead, and you risk paying for extra if you are not able to finish on time.

g. "Voluntary" price cuts: you intentionally reduce your overhead or profits to get the job. The eagerness is not a valid excuse. Construction is a business which should yield a reasonable profit.

FIGURE 5.2 — TIPS TO REDUCE ESTIMATING MISTAKES

- Be organized and keep your desk clean
- Get information from a complete set of drawings instead of a few sheets
- Read specs
- Use estimating checklists
- Use estimating forms
- Spend more time on large cost items
- Mark drawings when taking off items
- Prepare detailed estimates instead of estimating by square feet
- Figure material and labor for each item instead of applying a combined unit rate
- Round up the results in each step of calculation and drop the pennies
- Have someone else check your estimate and take-off
- Use specialized HVAC estimating software
- Check all formulas if using spreadsheet programs
- Compare costs with similar projects on a unit price basis
- Always verify site conditions with the drawings
- Ask questions instead of making assumptions
- Take your time and never rush the estimate

FIGURE 5.3 — POST-BID REVIEW WORKSHEET

Job Name:	Ref No.		Reviewer:
Type of Building:		Location:	
Proposal Amount: $			
Square Feet:		Cost per SF:	

Breakdown	Amount	% of Total Cost	Cost per SF
Equipment			
Fans and Blowers			
Pumps and Tanks			
Piping			
Ductwork			
Air Distribution Devices			
Insulation			
Instrumentation and Controls			
Electrical Wiring			
Air Balancing and System Testing			
Fastening			
Equipment Supports			
Rigging			
Miscellaneous			
Jobsite Overhead			
Office Overhead			
Owner's Allowance			
Contingency			
Total Cost			

FIGURE 5.4 – BID HISTORY TRACKING WORKSHEET

Job Name	Bid Date	Square Feet	Location	Job Type	Owner	General Contractor	Bid Result

FIGURE 5.5 — PREPARING A REVISED ESTIMATE FOR CONTRACT

1. Request two complete sets of latest drawings and specs (civil, architectural, structural, mechanical, electrical, etc.).

2. Review documents page by page to see what has changed since the bid. Write down the differences in detail.

3. Check the original list of inclusions, exclusions, and assumptions. See if there should be new issues added.

4. Track the original Request for Information (RFI) sent to the general contractor or design engineer. Send new RFI to ask more questions. Get all answers in writing (letter, fax, or e-mail). Only after an item has been clarified in writing to your satisfaction can it be removed from the inclusion and exclusion list.

5. Perform revised quantity take-off with greater accuracy. If necessary, start from scratch. Always keep the old and new estimates separate.

6. Get price updates from suppliers and subs. Make sure they have access to the latest design information. Prices might have changed dramatically since the bid.

7. Get a correct new total number and submit a revised proposal to the customer.

FIGURE 5.6 — ELEMENTS IN AN HVAC SUBCONTRACT

- Parties to the Contract
- Project Identification
- Scope of Work
- Schedule of Values
- Construction Schedule
- Submittals and Shop Drawings
- Quality
- Payment
- Change Orders
- Insurance, License, and Bond Requirements
- Warranty
- Protection of Work
- Safety
- Clean-up
- Termination
- Dispute Resolution
- General Clauses
- Attachments

FIGURE 5.7 — SCOPE OF WORK IN AN EXAMPLE HVAC SUBCONTRACT

The subcontractor (i.e. HVAC contractor) should provide all materials, equipment, tools, and supervision required to furnish and install all HVAC and all appurtenances required for a complete installation in accordance with the contract documents, applicable codes, and governing agencies. The work includes, but is not limited to, the following:

1. Furnish and install all equipment required to complete the HVAC system, including, but not limited to, furnace, fan coil, and condenser.

2. Furnish and install all required ducting, including, but not limited to, ducting, supply registers, return-air grilles, combustion air ducting, and weather caps.

3. Furnish and install ducting for clothes dryers.

4. Furnish and install ducting to all exhaust fans.

5. Furnish and install smoke and fire dampers as required.

6. Furnish and install thermostats and required low-voltage wiring, including temporary thermostats during construction (if applicable).

7. Furnish and install all refrigerant lines.

8. Furnish and install condensate drains lines and piping for HVAC equipment.

9. Furnish and install all natural gas lines including services to appliances such as ranges, dryers, fireplaces, water heaters, etc.

**FIGURE 5.7 — SCOPE OF WORK IN AN EXAMPLE
HVAC SUBCONTRACT** *(cont.)*

10. Connect all gas-powered appliances to gas lines after appliance installation.

11. The subcontractor will furnish all pipe flashings and sheet metal jacks to be installed by others.

12. The subcontractor will furnish all block-outs and sleeves through floors, walls, and ceilings required for its work.

13. The subcontractor will furnish and install fire stopping to seal all penetrations through fire-rated assemblies.

14. The subcontractor will coordinate with the steel subcontractor for location and size of roof openings for the HVAC equipment.

15. The subcontractor will coordinate with the framing subcontractor for location and size of all required backing for the HVAC equipment.

16. The subcontractor will work with the general contractor to provide an early operation of the furnace to be used as temporary heat in the building.

FIGURE 5.8 — NEGATIVE CLAUSES IN A BAD HVAC SUBCONTRACT

Watch for the following "No-No" traps (i.e. the clauses should be eliminated or revised):

✘ **"Pay If Paid":** Your payment from the general contractor depends on whether they receive payment from the owner first.

✘ **"No Damage for Delay":** You will be allowed no more than a time extension for the delay you did not cause, and absolutely no monetary compensation.

✘ **"Unconditional Lien Waiver before Payment":** You have to give up all of your lien and bond rights first, or they will not pay you.

✘ **"Change Order by Notice":** You must perform the additional work as directed by the general contractor without a written agreement in advance.

✘ **"Dispute Resolution by Litigation":** Any disputes will be resolved in the court of the general contractor's home state, which may be far from where the project is located.

✘ **"Broad Form Indemnity":** You are required to name general contractors or owners as "additional names insured" on your insurance policy.

✘ **"Subcontract Supercedes Prior Negotiation":** This contract will overrule your bid proposal or any other prior negotiations.

✘ **"Acceptance of Final Payment as Waiver":** You give up all your legal rights to existing unresolved claims when you receive all payment in full as stated by this subcontract.

FIGURE 5.9 — "TURN-OVER" MEETING AGENDA

1. Have a job overview. Everyone should know the job name and location, building square footage, name of client, design engineer, etc.

2. Review contract documents including drawings, specs, and addendums.

3. Review bid proposal and contract amount. What pricing assumptions were made?

4. Discuss site conditions and material storage problems.

5. Examine potential design problems, requests for information, and change orders.

6. Discuss project schedule and long lead items.

7. Review low subs, suppliers, and price escalation possibilities.

8. Evaluate risks. Make a list of make/break issues affecting the job costs.

CHAPTER 6
Man-hour Tables

Disclaimer: These man-hour tables represent the author's best judgment and care for the information published. Instructions should always be carefully studied before using the data. Neither the author nor the publisher is responsible for any losses or damages with respect to the accuracy, correctness, value, or sufficiency of the data contained herein.

There are many factors that influence how much work a man can finish within an hour. Factors such as jobsite conditions, supervision, tools and equipment, weather, code requirements, etc. can all have an impact on how efficiently work can be completed.

The man-hours tables in this chapter are based on an average craftsman working under normal conditions: new construction with fair productivity, standard materials and straight-forward installation, appropriate tools and good coordination with other trades.

Most of the man-hours listed are for hand labor only and provide general information on how long it normally takes to install a single unit of a certain item. They are based upon industry averages nationwide. You will need to evaluate specific job conditions and make adjustments.

UNDERSTANDING MAN-HOUR TABLES

How accurate are the man-hour tables in this chapter? For example: does it really take more than 3 hours to install a boiler?

The man-hours listed in these tables include:
- Unloading, storing, and getting raw materials
- Getting and returning tools/equipment
- Normal time lost due to work breaks
- Planning and discussing the work to be performed
- Normal handling, measuring, cutting, and fitting
- Layout of hangers and pipe support
- Test and balance of equipment
- Supervision time (foremen, superintendents, etc.)
- Setting up and tearing down of scaffoldings and ladders
- Regular cleanup of construction debris
- Infrequent correction or repairs required because of faulty installation

Please remember that in writing these tables there are unknowns such as:
- Specific experience or training of your crew
- Plans or specs of your target job
- Your job location and what building code applies to your job

ADJUSTING STANDARD MAN-HOURS

Estimating is an art, not a science. Installation times for plumbing items vary widely from job to job, from crew to crew, and even for the same crew from day to day. There is no one number that applies on all jobs. Use the following worksheet to adjust the standard man-hours in this chapter:

Working Conditions	Plus Percentage	Minus Percentage
Quality of Supervision		
Labor Situation		
Type of Work		
Inspections/Specs		
Working Hours		
Distance to Stocking Pile		
Overhead Piping		
High-Rise Building		
In Attic Space		
In Crawl Space/Tunnel		
In Shored Trench		
Special System		
Total Adjustment Percentage		

Estimating Example:

If it normally takes 3.00 hours to install a small residential boiler, but you are currently working on 20th floor and decide 25% needs to be added to the time.

Then the adjusted man-hour should be

$3.00 \times (100\% + 25\%) = 3.75$ hours for one boiler installation.

WRITING YOUR OWN MAN-HOUR TABLES

If you feel more comfortable writing your own man-hour tables, the following few pages will help you do it properly.

Estimating Math

Total Man-hours = Number of Working Crew Members $\times$ Hours Crew Worked

Unit Man-hour for the Item = Crew Output/Total Man-hours

Estimating Example

A crew of four spent one 8-hour day installing 1000 feet of ductwork.

Total man-hours are: 4 people $\times$ 8 hours = 32 hours

One linear foot of such ductwork will take: 32 hours /1000 feet = 0.032 hours/foot

When recording man-hour information, you can specify the following:

1. Materials and Installation Methods (e.g. dimensions and types of ductwork)
2. Types of jobs (e.g. new construction or renovation, residential or non-residential)
3. Crew (e.g. just one guy or a team including supervision and apprentice)
4. Tools and Equipment (e.g. scaffolding, lift, etc.)
5. Weather conditions (e.g. rain, snow, wind, and temperature)

Check your job at the same time each morning and record the number of units (such as the length of ductwork) installed the previous day. Repeat this on different projects over a period of time to verify your result.

JOBSITE DAILY MAN-HOUR WORKSHEET

Job Name:

Ref No.

Superintendent:

Date:

Sheet Numbers:

Item	Daily Output	Crew Member	Crew Hours	Total Man-Hours	Man-hours Per Unit

CONVERTING MINUTES TO DECIMAL HOURS

Minutes	Decimal in Hours	Minutes	Decimal in Hours
1	0.017	31	0.517
2	0.033	32	0.533
3	0.050	33	0.550
4	0.067	34	0.567
5	0.083	35	0.583
6	0.100	36	0.600
7	0.117	37	0.617
8	0.133	38	0.633
9	0.150	39	0.650
10	0.167	40	0.667
11	0.183	41	0.683
12	0.200	42	0.700
13	0.217	43	0.717
14	0.233	44	0.733
15	0.250	45	0.750
16	0.267	46	0.767
17	0.283	47	0.783
18	0.300	48	0.800
19	0.317	49	0.817
20	0.333	50	0.833
21	0.350	51	0.850
22	0.367	52	0.867
23	0.383	53	0.883
24	0.400	54	0.900
25	0.417	55	0.917
26	0.433	56	0.933
27	0.450	57	0.950
28	0.467	58	0.967
29	0.483	59	0.983
30	0.500	60	1.000

ROOF TOP UNITS

Size	Single Zone	Single Zone (VAV)	Multi Zone
3.0 Tons, 1,200 CFM	6.00	6.60	7.50
5.0 Tons, 2,000 CFM	8.00	8.80	10.00
7.5 Tons, 3,000 CFM	10.00	11.00	12.50
10.0 Tons, 4,000 CFM	10.00	11.00	12.50
12.5 Tons, 5,000 CFM	12.00	13.20	15.00
15.0 Tons, 6,000 CFM	12.00	13.20	15.00
20.0 Tons, 8,000 CFM	14.00	15.40	17.50
25.0 Tons, 10,000 CFM	14.00	15.40	17.50
30.0 Tons, 12,000 CFM	17.00	18.70	21.25
40.0 Tons, 16,000 CFM	22.00	24.20	27.50
50.0 Tons, 20,000 CFM	24.00	26.40	30.00
60.0 Tons, 24,000 CFM	29.00	31.90	36.25

AIR HANDLING UNITS

Size	Single Zone	Single Zone (VAV)	Multi Zone
3.0 Tons, 1,200 CFM	5.00	5.50	6.25
5.0 Tons, 2,000 CFM	9.00	9.90	1.25
7.5 Tons, 3,000 CFM	10.00	11.00	12.50
10.0 Tons, 4,000 CFM	13.00	14.30	16.20
12.5 Tons, 5,000 CFM	15.00	16.50	18.70
15.0 Tons, 6,000 CFM	17.00	18.70	21.20
20.0 Tons, 8,000 CFM	19.00	20.90	23.70
25.0 Tons, 10,000 CFM	21.00	23.10	26.20
30.0 Tons, 12,000 CFM	24.00	26.40	30.00
40.0 Tons, 16,000 CFM	28.00	30.80	35.00
50.0 Tons, 20,000 CFM	32.00	35.20	40.00
60.0 Tons, 24,000 CFM	36.00	39.60	45.00

SELF CONTAINED A.C. UNITS

Size	Single Zone	Single Zone (VAV)	Multi Zone
3.0 Tons, 1,200 CFM	6.00	6.60	7.50
5.0 Tons, 2,000 CFM	12.00	13.20	15.00
7.5 Tons, 3,000 CFM	14.00	15.40	17.50
10.0 Tons, 4,000 CFM	15.00	16.50	18.75
12.5 Tons, 5,000 CFM	18.00	19.80	22.50
15.0 Tons, 6,000 CFM	20.00	22.00	25.00
20.0 Tons, 8,000 CFM	22.00	24.20	27.50
25.0 Tons, 10,000 CFM	24.00	26.40	30.00
30.0 Tons, 12,000 CFM	26.00	28.60	32.50
40.0 Tons, 16,000 CFM	31.00	34.10	38.75
50.0 Tons, 20,000 CFM	36.00	39.60	45.50
60.0 Tons, 24,000 CFM	42.00	46.20	52.50

AIR HANDLING UNIT ACCESSORIES

Variable Speed Drive

Item	Hour per Drive
5 HP	4.00
7.5 HP	4.00
10 HP	6.00
15 HP	8.00
20 HP	12.00
25 HP	16.00
30 HP	18.00
40 HP	22.00
50 HP	26.00

Variable Inlet Vane

Item	Hour per Drive
1,000 to 1,500 CFM	4.00
1,600 to 2,500 CFM	4.00
2,600 to 5,000 CFM	6.00
6,000 to 10,000 CFM	8.00
11,000 to 20,000 CFM	10.00

AIR HANDLING UNIT ACCESSORIES *(cont.)*

Coil Connection — One Row Coil Bank

Item	Non-Regulated Flow Design	2-Way Control Valve Design	3-Way Control Valve Design
1½" supply	14.0	16.5	18.0
2" supply	18.0	21.0	26.0
2½" supply	28.0	33.6	41.5
3" supply	31.5	36.0	43.8
4" supply	34.0	43.8	54.6
6" supply	41.0	54.0	61.4

Coil Connection — Two Row Coil Bank

Item	Non-Regulated Flow Design	2-Way Control Valve Design	3-Way Control Valve Design
2½" supply	48.0	54.0	58.8
3" supply	51.5	58.5	64.0
4" supply	63.6	67.0	72.0
6" supply	79.2	83.5	87.0

Coil Connection — Three Row Coil Bank

Item	Non-Regulated Flow Design	2-Way Control Valve Design	3-Way Control Valve Design
4" supply	72.0	78.0	82.0
6" supply	94.0	105.0	112.0
8" supply	120.0	132.0	140.0

CONDENSING UNITS		
Size	Air Cooled	Water Cooled
3 Tons	6.00	8.00
5 Tons	8.00	8.00
7½ Tons	8.00	12.00
10 Tons	12.00	12.00
15 Tons	16.00	20.00
20 Tons	16.00	20.00
30 Tons	24.00	24.00
40 Tons	24.00	24.00
50 Tons	24.00	32.00
60 Tons	32.00	32.00
75 Tons	32.00	32.00
100 Tons	40.00	32.00

EVAPORATIVE CONDENSERS	
Size	**Hours per Condenser**
7½ Tons	7.00
10 Tons	8.50
15 Tons	11.50
20 Tons	15.00
25 Tons	16.00
30 Tons	18.50
40 Tons	23.00
50 Tons	29.00
60 Tons	32.00

COMPRESSORS	
Size	**Hours per Compressor**
3 Tons	8.00
5 Tons	8.00
7½ Tons	8.00
10 Tons	12.00
15 Tons	12.0*0
20 Tons	16.00
30 Tons	16.00
40 Tons	16.00
50 Tons	16.00
60 Tons	32.00
75 Tons	32.00
100 Tons	36.00
125 Tons	40.00
150 Tons	48.00

COOLING TOWERS	
Cooling Tower – Galvanized	
Size	**Hours per Cooling Tower**
10 Tons	16.00
15 Tons	16.00
20 Tons	18.00
25 Tons	20.00
30 Tons	25.00
40 Tons	30.00
50 Tons	35.00
60 Tons	40.00
80 Tons	45.00
100 Tons	50.00
125 Tons	60.00
150 Tons	65.00
175 Tons	70.00
200 Tons	70.00
300 Tons	75.00
400 Tons	80.00
500 Tons	80.00
Cooling Tower – Redwood/Fir Induced Draft	
Size	**Hours per Cooling Tower**
100 Tons	75.00
200 Tons	80.00
300 Tons	85.00
400 Tons	90.00
500 Tons	90.00

CHILLERS	
Chiller – Centrifugal	
Size	**Hours per Chiller**
100 Tons	72.00
200 Tons	72.00
300 Tons	80.00
400 Tons	80.00
500 Tons	80.00
Chiller – Reciprocating	
Size	**Hours per Chiller**
10 Tons	16.00
15 Tons	16.00
20 Tons	16.00
25 Tons	20.00
30 Tons	20.00
40 Tons	20.00
50 Tons	20.00
60 Tons	24.00
75 Tons	24.00
100 Tons	27.00
125 Tons	32.00
150 Tons	38.00

CHILLERS *(cont.)*	
Chiller – Absorption	
Size	**Hours per Chiller**
75 Tons	64.00
100 Tons	64.00
150 Tons	64.00
300 Tons	80.00
500 Tons	80.00
700 Tons	120.00
1000 Tons	120.00
Chiller – Absorption (Residential/Light Commercial)	
Size	**Hours per Chiller**
3 Tons	10.00
5 Tons	10.00
7.5 Tons	10.00
15 Tons	12.00
25 Tons	16.00

CHILLED WATER COILS	
Chilled Water Coil — Two Rows	
Size	**Hours per Coil**
12" × 12"	1.80
18" × 12"	2.00
24" × 12"	2.20
24" × 24"	2.60
36" × 18"	3.00
36" × 24"	3.60
48" × 24"	5.00
Chilled Water Coil — Six Rows	
Size	**Hours per Coil**
36" × 18"	4.00
36" × 24"	6.00
48" × 24"	7.00
48" × 36"	10.00
60" × 48"	14.00
72" × 60"	18.00
96" × 60"	22.00
108" × 72"	23.00
120" × 72"	27.00

WINDOW A/C UNITS	
Installation Condition	**Hours per Unit**
Correct size precast opening	3.00
Enlarge a precast opening	5.00
Concrete block wall	8.00
Jalousie window	5.00
Casement window	8.00
Awning window	7.00
Double-hung window	4.00
Transom	6.00

BOILERS	
Residential Boiler	
Size	**Hours per Boiler**
3.0 HP, 100 MBH	2.0
4.5 HP, 150 MBH	2.5
6.0 HP, 200 MBH	3.0
Miscellaneous	
Item	**Hours per Item**
Conversion Burner	6.0
Circulator	0.5
Zone/Control Valve	0.5
Expansion Tank	1.0
Humidifier	2.0
Filter	2.0
Wiring	2.5
Service/Balancing	4.0

BOILERS *(cont.)*	
Cast Iron Boiler	
Size	**Hours per Boiler**
4.5 HP, 150 MBH	20.0
6.0 HP, 200 MBH	25.0
9.0 HP, 300 MBH	32.0
15.0 HP, 500 MBH	42.0
20.0 HP, 670 MBH	46.0
22.5 HP, 750 MBH	50.0
30.0 HP, 1000 MBH	64.0
45.0 HP, 1500 MBH	80.0
60.0 HP, 2000 MBH	114.0
82.0 HP, 2750 MBH	135.0
104.5 HP, 3500 MBH	180.0
127.0 HP, 4250 MBH	200.0
150.0 HP, 5000 MBH	240.0
180.0 HP, 6000 MBH	280.0
225.0 HP, 7500 MBH	310.0

BOILERS *(cont.)*	
Firetube Boiler	
Size	**Hours per Boiler**
30 HP, 1000 MBH	16.0
40 HP, 1340 MBH	24.0
50 HP, 1670 MBH	24.0
60 HP, 2010 MBH	28.0
75 HP, 2510 MBH	32.0
80 HP, 2680 MBH	36.0
100 HP, 3350 MBH	40.0
150 HP, 5020 MBH	40.0
200 HP, 6700 MBH	64.0
Watertube Gas Boiler	
Size	**Hours per Boiler**
4.5 HP, 150 MBH	8.0
9.0 HP, 300 MBH	12.0
15.0 HP, 500 MBH	12.0
22.5 HP, 750 MBH	16.0
30.0 HP, 1000 MBH	16.0
45.0 HP, 1500 MBH	24.0
60.0 HP, 2000 MBH	24.0
82.0 HP, 2750 MBH	24.0
104.5 HP, 3500 MBH	40.0
127.0 HP, 4250 MBH	40.0
150.0 HP, 5000 MBH	40.0
180.0 HP, 6000 MBH	64.0
225.0 HP, 7500 MBH	64.0

BOILERS *(cont.)*	
Watertube Oil Boiler	
Size	**Hours per Boiler**
4.5 HP, 150 MBH	10.0
9.0 HP, 300 MBH	16.0
14.9 HP, 500 MBH	16.0
22.4 HP, 750 MBH	20.0
29.9 HP, 1000 MBH	20.0
44.8 HP, 1500 MBH	30.0
59.7 HP, 2000 MBH	30.0
82.1 HP, 2750 MBH	30.0
104.5 HP, 3500 MBH	48.0
126.9 HP, 4250 MBH	48.0
149.3 HP, 5000 MBH	48.0
179.2 HP, 6000 MBH	72.0
224.0 HP, 7500 MBH	72.0

BOILER ACCESSORIES	
Burner	
Size	**Hours per Burner**
300 MBH	8.0
500 MBH	8.0
800 MBH	12.0
1000 MBH	12.0
Deaerator/Condenser Unit	
Boiler Size	**Hours per Pump**
100 HP	8.0
200 HP	12.0
300 HP	14.0
400 HP	16.0
600 HP	24.0
800 HP	36.0
1000 HP	48.0
Stack Waste Heat Recovery System	
Boiler Size	**Hours per System**
500 HP	40.0
750 HP	48.0
800 HP	48.0
1000 HP	56.0

BOILER ACCESSORIES (cont.)

Feedwater Pump

Boiler Size	Hours per Pump
13 to 100 HP	8.0
125 to 200 HP	16.0
250 to 800 HP	32.0

Chemical Feed Duplex Pump

Boiler Size	Hours per Pump
20 to 100 HP	3.0
125 to 200 HP	3.0
200 to 400 HP	6.0
500 to 600 HP	6.0
700 to 800 HP	8.0
1,000 HP	10.0

Combustion Control

Boiler Size	Hours per Control
13 to 100 HP	4.0
125 to 200 HP	8.0
250 to 400 HP	16.0
500 to 600 HP	32.0
750 to 800 HP	64.0

BOILER ACCESSORIES (cont.)	
Water Softening System	
Boiler Size	**Hours per System**
13 to 100 HP	4.0
125 to 250 HP	8.0
300 to 500 HP	16.0
600 HP	32.0
750 to 800 HP	64.0
Boiler Trim	
Boiler Size	**Hours per Trim**
20 to 100 HP	4.0
100 to 200 HP	6.0
200 to 400 HP	8.0
400 to 600 HP	8.0
600 to 800 HP	8.0
800 to 1000 HP	10.0
Start-up, Shakedown, and Calibration	
Boiler Size	**Hours**
13 to 100 HP	4.0
125 to 200 HP	8.0
250 to 400 HP	16.0
500 to 600 HP	24.0
750 to 800 HP	30.0

FURNACES	
Size	Hours per Furnace
30 MBH	2.5
50 MBH	2.8
70 MBH	3.0
100 MBH	3.2
150 MBH	3.5
200 MBH	4.0

UNIT HEATERS	
Hot Water Unit Heater	
Size	**Hours per Heater**
16 MBH	2.0
30 MBH	3.0
43 MBH	3.5
57 MBH	4.0
68 MBH	4.0
105 MBH	5.0
123 MBH	6.0
200 MBH	6.0
250 MBH	8.0
300 MBH	8.0
Gas Fired Unit Heater	
Size	**Hours per Heater**
25 MBH	5.0
50 MBH	6.0
75 MBH	6.0
100 MBH	7.0
150 MBH	8.0
200 MBH	9.0
250 MBH	10.0
300 MBH	11.0
350 MBH	12.0
400 MBH	13.0

DUCT HEATERS	
Size	**Hours per Heater**
25 MBH	4.0
50 MBH	5.0
75 MBH	5.0
100 MBH	6.0
150 MBH	7.0
200 MBH	8.0
250 MBH	9.0
300 MBH	10.0
350 MBH	11.0
400 MBH	12.0
500 MBH	12.0
INFRARED HEATERS	
Size	**Hours per Heater**
30 MBH	1.2
60 MBH	1.8
100 MBH	2.5
150 MBH	2.8
175 MBH	2.8
200 MBH	3.0

FAN COIL UNITS	
Fan Coil Unit	
Size	**Hours per Unit**
200 CFM	2.15
400 CFM	2.75
600 CFM	3.25
800 CFM	3.45
1000 CFM	3.60
Fan Coil Unit Connection	
Size	**Hours per Connection**
1/2" Supply	2.75
3/4" Supply	2.95
1" Supply	3.40

REHEAT COILS		
Reheat Coil		
Size	**Hot Water**	**Electric**
12" × 6"	1.00	1.10
12" × 8"	1.00	1.20
12" × 10"	1.00	1.30
12" × 12"	1.00	1.40
18" × 6"	1.25	1.50
18" × 12"	1.25	1.60
18" × 18"	1.25	1.70
24" × 12"	1.50	1.90
24" × 18"	1.50	2.10
24" × 24"	1.75	2.40
Reheat Coil Connection		
Size	**Hours per Connection**	
½" Supply	2.25	
¾" Supply	2.35	
1" Supply	2.65	
1¼" Supply	3.00	
1½" Supply	3.25	
2" Supply	3.75	

WATER HEATERS	
Item	**Hours per Heater**
20 Gallons	2.75
30 Gallons	3.00
40 Gallons	4.00
50 Gallons	4.50
60 Gallons	5.00
80 Gallons	5.50
120 Gallons	6.50
240 Gallons	8.00
360 Gallons	8.00
480 Gallons	10.00
840 Gallons	12.00
920 Gallons	16.00
1,200 Gallons	20.00
HEAT PUMPS	
Item	**Hours per Pump**
1.5 Tons	4.00
2.0 Tons	4.25
2.5 Tons	4.50
3.0 Tons	5.00
3.5 Tons	5.75
4.0 Tons	6.25
5.0 Tons	7.50

HEAT EXCHANGERS

Heat Exchanger

Size	Hours per Item
4" Diameter × 30" long	2.00
4" Diameter × 60" long	2.50
6" Diameter × 30" long	3.25
6" Diameter × 60" long	3.50
8" Diameter × 30" long	4.25
8" Diameter × 60" long	4.50
10" Diameter × 30" long	5.50
10" Diameter × 60" long	5.75
10" Diameter × 90" long	6.00
12" Diameter × 60" long	6.50
12" Diameter × 90" long	7.00
12" Diameter × 120" long	7.25
14" Diameter × 60" long	7.50
14" Diameter × 90" long	8.00
14" Diameter × 120" long	8.25
16" Diameter × 60" long	8.50
16" Diameter × 90" long	9.00
16" Diameter × 120" long	9.25

Heat Exchanger Connections

Item	Hot Water to Hot Water	Steam to Hot Water
1½" supply	32.0	22.0
2" supply	38.0	26.0
2½" supply	50.0	41.0
3" supply	55.0	44.0
4" supply	62.0	54.5
6" supply	85.0	67.0
8" supply	102.0	80.0
10" supply	130.0	N/A

HEAT RECOVERY VENTILATORS		
Residential Heat Recovery Ventilator		
Size	**Hours per Ventilator**	
65 to 127 CFM	2.65	
115 to 180 CFM	2.90	
180 to 265 CFM	3.25	
Commercial Heat Recovery Ventilator		
Size	**Hours per Ventilator**	
700 CFM	5.00	
1200 CFM	5.75	
2500 CFM	6.75	
UNIT VENTILATORS		
Size	**Floor**	**Suspended**
50 MBH	5.0	6.0
75 MBH	6.0	7.0
100 MBH	9.0	10.5
150 MBH	12.0	13.5
200 MBH	14.0	16.0

FANS		
Centrifugal Fans		
Size	Regular	Suspended
1,000 CFM	3.00	3.30
2,000 CFM	7.00	7.70
4,000 CFM	10.00	11.00
6,000 CFM	12.00	13.20
8,000 CFM	14.00	15.40
10,000 CFM	16.00	17.60
12,000 CFM	18.00	19.80
14,000 CFM	20.00	22.00
16,000 CFM	22.00	24.20
18,000 CFM	24.00	26.40
20,000 CFM	26.00	28.60
25,000 CFM	30.00	33.00
30,000 CFM	36.00	39.60
40,000 CFM	46.00	50.60
50,000 CFM	56.00	61.60
60,000 CFM	64.00	70.40

FANS *(cont.)*		
Centrifugal Fans — Utility Set		
Size	**Regular**	**Suspended**
500 CFM	2.00	2.20
1,000 CFM	4.00	4.40
2,000 CFM	5.00	5.50
4,000 CFM	7.00	7.70
6,000 CFM	9.00	9.90
8,000 CFM	10.00	11.00
10,000 CFM	12.00	13.20
12,000 CFM	14.00	15.40
14,000 CFM	16.00	17.60
16,000 CFM	18.00	19.80
18,000 CFM	20.00	22.00
20,000 CFM	22.00	24.20
Vane-Axial Fans		
Size	**Regular**	**Suspended**
18,000 CFM	20.00	22.00
20,000 CFM	24.00	26.40
25,000 CFM	28.00	30.80
30,000 CFM	32.00	35.20
40,000 CFM	36.00	39.60
50,000 CFM	46.00	50.60
60,000 CFM	50.00	55.00
80,000 CFM	54.00	59.40
100,000 CFM	64.00	70.40

FANS *(cont.)*	
Roof Exhaust Fans	
Size	**Hours per Fan**
500 CFM	3.00
1,000 CFM	3.00
2,000 CFM	4.00
4,000 CFM	5.00
6,000 CFM	6.00
8,000 CFM	6.00
10,000 CFM	7.00
12,000 CFM .	7.00
14,000 CFM	8.00
16,000 CFM	8.00
18,000 CFM	9.00
20,000 CFM	9.00
Ceiling Exhaust Fans	
Size	**Hours per Fan**
90 CFM	1.00
100 CFM	1.05
150 CFM	1.15
151 CFM	1.25
200 CFM	1.35
250 CFM	1.45
270 CFM	1.50
300 CFM	1.65
400 CFM	2.15
500 CFM	2.50
700 CFM	2.75
900 CFM	2.85
1,500 CFM	3.50
2,000 CFM	3.85
3,500 CFM	4.25

CENTRIFUGAL PUMPS	
Pump	
Item	**Hour per Pump**
⅙ HP	1.25
¼ HP	1.40
⅓ HP	1.50
½ HP	1.70
¾ HP	1.80
1 HP	2.00
1½ HP	2.40
2 HP	2.70
3 HP	3.00
5 HP	3.50
7½ HP	4.00
10 HP	4.50
15 HP	5.00
20 HP	6.00
25 HP	7.00
30 HP	7.40
40 HP	7.70
50 HP	8.00
Pump Connection	
Item	**Hours per Connection**
2½" supply	12.0
3" supply	13.5
4" supply	16.0
6" supply	22.0
8" supply	28.0
10" supply	35.0
12" supply	42.0

TANKS	
Residential Storage Tank	
Item	**Hours per Tank**
55 Gallons	1.00
112 Gallons	1.20
280 Gallons	2.00
560 Gallons	2.80
Commercial Storage Tank	
Item	**Hours per Tank**
1,000 Gallons	5.20
1,100 Gallons	5.60
1,500 Gallons	6.40
2,000 Gallons	8.00
2,500 Gallons	9.20
3,000 Gallons	12.00
4,000 Gallons	18.00
5,000 Gallons	23.00
7,500 Gallons	25.50
8,500 Gallons	30.00
10,000 Gallons	34.00
12,000 Gallons	42.00
15,000 Gallons	48.00
20,000 Gallons	62.00
25,000 Gallons	90.00
30,000 Gallons	112.00

TANKS *(cont.)*	
Hot Water Generator Tank	
Item	Hours per Tank
53 Gallons	2.60
120 Gallons	2.80
140 Gallons	3.00
220 Gallons	3.40
250 Gallons	4.10
315 Gallons	4.60
420 Gallons	5.00
520 Gallons	6.30
575 Gallons	6.80
720 Gallons	7.40
750 Gallons	7.60
940 Gallons	9.20
1,000 Gallons	10.30
1,130 Gallons	11.20
1,425 Gallons	14.00
1,760 Gallons	17.20
2,000 Gallons	18.20
2,050 Gallons	20.00
3,000 Gallons	28.00
4,000 Gallons	40.00

TANKS *(cont.)*	
Expansion Tank	
Item	**Hours per Tank**
15 Gallons	2.00
25 Gallons	2.10
30 Gallons	2.30
40 Gallons	2.50
60 Gallons	2.60
80 Gallons	2.80
100 Gallons	3.00
120 Gallons	3.20
140 Gallons	3.50
180 Gallons	3.70
CONTROLS	
Item	**Hours per Item**
Thermostat	1.00
Sensor	0.70
Controller	1.40
Relay	0.80
Switch	0.55
Motor	1.20
Valve	0.80

GALVANIZED STEEL DUCTWORK
(HOURS PER POUND)

Low Pressure Ductwork

Item	Fabrication	Erection	Fabrication & Erection
26 Gauge	0.067	0.067	0.134
24 Gauge	0.050	0.053	0.103
22 Gauge	0.042	0.043	0.085
20 Gauge	0.035	0.037	0.072
18 Gauge	0.033	0.034	0.067

Medium/High Ductwork

Item	Fabrication	Erection	Fabrication & Erection
26 Gauge	0.071	0.071	0.142
24 Gauge	0.056	0.059	0.115
22 Gauge	0.045	0.048	0.093
20 Gauge	0.040	0.042	0.082
18 Gauge	0.037	0.038	0.075

FIBERGLASS DUCTWORK (HOURS PER FOOT)

Type 475 Board

Width + Depth	Fabrication	Erection	Fabrication & Erection
12"	0.046	0.070	0.116
14"	0.052	0.082	0.134
16"	0.058	0.093	0.151
18"	0.066	0.105	0.171
20"	0.074	0.117	0.191
22"	0.080	0.128	0.208
24"	0.088	0.140	0.228
26"	0.096	0.152	0.248
28"	0.102	0.163	0.265
30"	0.110	0.175	0.285
32"	0.128	0.187	0.315
34"	0.136	0.198	0.334
36"	0.144	0.210	0.354
38"	0.152	0.222	0.374
40"	0.160	0.232	0.392
42"	0.168	0.245	0.413
44"	0.176	0.257	0.433

FIBERGLASS DUCTWORK
(HOURS PER FOOT) *(cont.)*

Type 475 Board

Width + Depth	Fabrication	Erection	Fabrication & Erection
46"	0.184	0.268	0.452
48"	0.192	0.280	0.472
50"	0.200	0.292	0.492
52"	0.208	0.303	0.511
54"	0.216	0.315	0.531
56"	0.234	0.327	0.561
58"	0.241	0.338	0.579
60"	0.250	0.350	0.600
64"	0.265	0.373	0.638
68"	0.285	0.397	0.682
72"	0.300	0.420	0.720
76"	0.315	0.443	0.758
80"	0.335	0.467	0.802
84"	0.350	0.490	0.840
88"	0.526	0.513	1.039
92"	0.554	0.537	1.091
96"	0.576	0.560	1.136

FIBERGLASS DUCTWORK
(HOURS PER FOOT) *(cont.)*

Type 800 Standard Duty Board

Width + Depth	Fabrication	Erection	Fabrication & Erection
12"	0.052	0.070	0.122
14"	0.056	0.082	0.138
16"	0.064	0.093	0.157
18"	0.072	0.105	0.177
20"	0.080	0.117	0.197
22"	0.088	0.128	0.216
24"	0.096	0.140	0.236
26"	0.104	0.152	0.256
28"	0.112	0.163	0.275
30"	0.120	0.175	0.295
32"	0.144	0.187	0.331
34"	0.153	0.198	0.351
36"	0.162	0.210	0.372
38"	0.171	0.222	0.393
40"	0.180	0.232	0.412
42"	0.189	0.245	0.434
44"	0.198	0.257	0.455

FIBERGLASS DUCTWORK
(HOURS PER FOOT) *(cont.)*

Type 800 Standard Duty Board

Width + Depth	Fabrication	Erection	Fabrication & Erection
46"	0.207	0.268	0.475
48"	0.216	0.280	0.496
50"	0.225	0.292	0.517
52"	0.234	0.303	0.537
54"	0.243	0.315	0.558
56"	0.262	0.327	0.589
58"	0.270	0.338	0.608
60"	0.280	0.350	0.630
64"	0.297	0.373	0.670
68"	0.319	0.397	0.716
72"	0.336	0.420	0.756
76"	0.353	0.443	0.796
80"	0.375	0.467	0.842
84"	0.392	0.490	0.882
88"	0.584	0.513	1.097
92"	0.616	0.537	1.153
96"	0.640	0.560	1.200

FIBERGLASS DUCTWORK
(HOURS PER FOOT) *(cont.)*

Type 800 Standard Duty Board with Hand Grooving

Width + Depth	Fabrication	Erection	Fabrication & Erection
12"	0.076	0.070	0.146
14"	0.087	0.082	0.169
16"	0.098	0.093	0.191
18"	0.111	0.105	0.216
20"	0.124	0.117	0.241
22"	0.135	0.128	0.263
24"	0.148	0.140	0.288
26"	0.161	0.152	0.313
28"	0.172	0.163	0.335
30"	0.185	0.175	0.360
32"	0.209	0.187	0.396
34"	0.220	0.198	0.418
36"	0.234	0.210	0.444
38"	0.248	0.222	0.470
40"	0.259	0.232	0.491
42"	0.273	0.245	0.518
44"	0.287	0.257	0.544

FIBERGLASS DUCTWORK
(HOURS PER FOOT) *(cont.)*

Type 800 Standard Duty Board with Hand Grooving

Width + Depth	Fabrication	Erection	Fabrication & Erection
46"	0.298	0.268	0.566
48"	0.312	0.280	0.592
50"	0.326	0.292	0.618
52"	0.337	0.303	0.640
54"	0.351	0.315	0.666
56"	0.393	0.327	0.720
58"	0.405	0.338	0.743
60"	0.420	0.350	0.770
64"	0.445	0.373	0.818
68"	0.479	0.397	0.876
72"	0.504	0.420	0.924
76"	0.529	0.443	0.972
80"	0.563	0.467	1.030
84"	0.588	0.490	1.078
88"	0.876	0.513	1.389
92"	0.924	0.537	1.461
96"	0.960	0.560	1.520

FIBERGLASS DUCTWORK
(HOURS PER FOOT) (cont.)

Type 1400 Board

Width + Depth	Fabrication	Erection	Fabrication & Erection
12"	0.062	0.070	0.132
14"	0.068	0.082	0.150
16"	0.077	0.093	0.170
18"	0.087	0.105	0.192
20"	0.097	0.117	0.214
22"	0.106	0.128	0.234
24"	0.116	0.140	0.256
26"	0.126	0.152	0.278
28"	0.135	0.163	0.298
30"	0.145	0.175	0.320
32"	0.171	0.187	0.358
34"	0.181	0.198	0.379
36"	0.192	0.210	0.402
38"	0.203	0.222	0.425
40"	0.213	0.232	0.445
42"	0.224	0.245	0.469
44"	0.235	0.257	0.492

FIBERGLASS DUCTWORK
(HOURS PER FOOT) *(cont.)*

Type 1400 Board

Width + Depth	Fabrication	Erection	Fabrication & Erection
46"	0.245	0.268	0.513
48"	0.256	0.280	0.536
50"	0.267	0.292	0.559
52"	0.277	0.303	0.580
54"	0.288	0.315	0.603
56"	0.309	0.327	0.636
58"	0.318	0.338	0.656
60"	0.330	0.350	0.680
64"	0.350	0.373	0.723
68"	0.376	0.397	0.773
72"	0.396	0.420	0.816
76"	0.416	0.443	0.859
80"	0.442	0.467	0.909
84"	0.462	0.490	0.952
88"	0.701	0.513	1.214
92"	0.739	0.537	1.276
96"	0.768	0.560	1.328

FLEX DUCT (HOURS PER FOOT)

Diameter	Fabrication & Erection
6"	0.08
7"	0.10
8"	0.13
9"	0.15
10"	0.18
12"	0.20
14"	0.25
16"	0.25
18"	0.30
20"	0.30
22"	0.35
24"	0.35

OTHER DUCTWORK (HOURS PER POUND)

Rectangular Black Iron Ductwork (Butt Welded Connection)

Item	Fabrication	Erection	Fabrication & Erection
18 Gauge	0.022	0.038	0.060
16 Gauge	0.019	0.034	0.053
14 Gauge	0.017	0.029	0.046
12 Gauge	0.014	0.026	0.040
10 Gauge	0.011	0.023	0.034
3/16 Inch	0.011	0.031	0.042
1/4 Inch	0.011	0.033	0.044

Rectangular Black Iron Ductwork (Coupling Connection)

Item	Fabrication	Erection	Fabrication & Erection
18 Gauge	0.036	0.029	0.065
16 Gauge	0.032	0.026	0.058
14 Gauge	0.029	0.022	0.051
12 Gauge	0.023	0.020	0.043
10 Gauge	0.019	0.018	0.037
3/16 Inch	0.019	0.024	0.043
1/4 Inch	0.018	0.025	0.043

OTHER DUCTWORK (HOURS PER POUND) *(cont.)*

Round Black Iron Ductwork

Item	Fabrication	Erection	Fabrication & Erection
18 Gauge	0.038	0.029	0.067
16 Gauge	0.034	0.026	0.060
14 Gauge	0.030	0.022	0.052
12 Gauge	0.024	0.020	0.044
10 Gauge	0.020	0.018	0.038
3/16 Inch	0.020	0.024	0.044
1/4 Inch	0.019	0.025	0.044

Stainless Steel Ductwork

Item	Fabrication	Erection	Fabrication & Erection
18 Gauge	0.045	0.036	0.081
16 Gauge	0.040	0.033	0.073
14 Gauge	0.036	0.028	0.064
12 Gauge	0.029	0.025	0.054
10 Gauge	0.024	0.022	0.046
3/16 Inch	0.024	0.030	0.054
1/4 Inch	0.022	0.031	0.053

Aluminum Ductwork

Item	Fabrication	Erection	Fabrication & Erection
18 Gauge	0.063	0.022	0.085
16 Gauge	0.056	0.020	0.076
14 Gauge	0.051	0.017	0.068
12 Gauge	0.040	0.015	0.055
10 Gauge	0.033	0.014	0.047
3/16 Inch	0.033	0.018	0.051
1/4 Inch	0.032	0.019	0.051

GALVANIZED STEEL RECTANGULAR DUCTWORK

26 Gauge Rectangular Ductwork (Hours per Foot)

Long Dimension	Short Dimension								
	4"	5"	6"	7"	8"	9"	10"	11"	12"
4"	0.048	N/A	N/A	N/A	N/A	N/A	N/A	N/A	N/A
5"	0.054	0.060	N/A	N/A	N/A	N/A	N/A	N/A	N/A
6"	0.060	0.066	0.072	N/A	N/A	N/A	N/A	N/A	N/A
7"	0.066	0.072	0.078	0.084	N/A	N/A	N/A	N/A	N/A
8"	0.072	0.078	0.084	0.091	0.097	N/A	N/A	N/A	N/A
9"	0.078	0.084	0.091	0.097	0.103	0.109	N/A	N/A	N/A
10"	0.084	0.091	0.097	0.103	0.109	0.115	0.129	N/A	N/A
11"	0.091	0.097	0.103	0.109	0.115	0.121	0.127	0.133	N/A
12"	0.097	0.103	0.109	0.115	0.121	0.127	0.133	0.139	0.145

GALVANIZED STEEL RECTANGULAR DUCTWORK (cont.)

24 Gauge Rectangular Ductwork (Hours per Foot)

One Side Dimension	One Side Dimension								
	14"	16"	18"	20"	22"	24"	26"	28"	30"
6"	0.154	0.170	0.184	0.200	0.216	0.231	0.247	0.262	0.278
8"	0.170	0.184	0.200	0.216	0.231	0.247	0.262	0.278	0.292
10"	0.184	0.200	0.216	0.231	0.247	0.262	0.278	0.292	0.308
12"	0.200	0.216	0.231	0.247	0.262	0.278	0.292	0.308	0.323
14" – 16"	0.223	0.239	0.254	0.269	0.285	0.300	0.316	0.331	0.346
18" – 20"	0.254	0.269	0.285	0.300	0.316	0.331	0.346	0.362	0.377
22" – 24"	0.285	0.300	0.316	0.331	0.346	0.362	0.377	0.392	0.407
26" – 28"	0.316	0.331	0.346	0.362	0.377	0.392	0.407	0.423	0.440
30"	0.339	0.354	0.370	0.385	0.400	0.415	0.431	0.448	0.465

GALVANIZED STEEL RECTANGULAR DUCTWORK (cont.)

22 Gauge Rectangular Ductwork (Hours per Foot)

One Side Dimension	One Side Dimension											
	32"	34"	36"	38"	40"	42"	44"	46"	48"	50"	52"	54"
8"	0.374	0.393	0.412	0.431	0.453	0.470	0.487	0.508	0.525	0.542	0.564	0.581
10"	0.393	0.412	0.431	0.453	0.470	0.487	0.508	0.525	0.542	0.564	0.581	0.598
12"	0.412	0.431	0.453	0.470	0.487	0.508	0.525	0.542	0.564	0.581	0.598	0.619
14" – 16"	0.442	0.461	0.478	0.497	0.517	0.534	0.555	0.572	0.589	0.611	0.628	0.645
18" – 20"	0.478	0.497	0.517	0.534	0.553	0.572	0.589	0.611	0.628	0.645	0.662	0.683
22" – 24"	0.517	0.534	0.553	0.572	0.589	0.611	0.628	0.649	0.664	0.683	0.700	0.722
26" – 28"	0.553	0.572	0.589	0.608	0.628	0.645	0.666	0.683	0.702	0.722	0.739	0.758
30" – 32"	0.589	0.608	0.628	0.645	0.664	0.683	0.705	0.722	0.739	0.758	0.777	0.796
34" – 36"	0.628	0.645	0.664	0.683	0.702	0.722	0.739	0.758	0.777	0.798	0.816	0.833
38" – 40"	0.664	0.683	0.702	0.722	0.739	0.758	0.777	0.796	0.816	0.833	0.852	0.871
42" – 44"	0.702	0.722	0.739	0.758	0.777	0.798	0.816	0.833	0.852	0.871	0.890	0.909
46" – 48"	0.739	0.758	0.777	0.796	0.816	0.833	0.852	0.871	0.890	0.909	0.927	0.946
50" – 52"	0.777	0.796	0.816	0.833	0.852	0.871	0.890	0.909	0.927	0.946	0.976	0.982
54"	0.807	0.824	0.841	0.863	0.880	0.901	0.918	0.935	0.956	0.974	0.991	1.010

GALVANIZED STEEL RECTANGULAR DUCTWORK *(cont.)*

20 Gauge Rectangular Ductwork (Hours per Foot)

One Side Dimension	One Side Dimension									
	56"	58"	60"	62"	64"	66"	68"	70"	72"
18" – 20"	0.828	0.850	0.871	0.895	0.916	0.937	0.959	0.980	1.000
22" – 24"	0.871	0.895	0.916	0.937	0.959	0.980	1.000	1.020	1.050
26" – 28"	0.916	0.937	0.959	0.980	1.000	1.020	1.050	1.070	1.090
30" – 32"	0.959	0.980	1.000	1.020	1.050	1.070	1.090	1.110	1.140
34" – 36"	1.000	1.020	1.050	1.070	1.090	1.110	1.140	1.160	1.180
38" – 40"	1.050	1.070	1.090	1.110	1.140	1.160	1.180	1.200	1.230
42" – 44"	1.090	1.110	1.140	1.160	1.180	1.200	1.230	1.250	1.270
46" – 48"	1.140	1.160	1.180	1.200	1.230	1.250	1.270	1.290	1.310
50" – 52"	1.180	1.020	1.230	1.250	1.270	1.290	1.310	1.330	1.360
54" – 56"	1.230	1.250	1.270	1.290	1.310	1.330	1.360	1.380	1.400
58" – 60"	1.270	1.290	1.310	1.330	1.360	1.380	1.400	1.420	1.450
62" – 64"	1.310	1.330	1.360	1.380	1.400	1.420	1.450	1.470	1.490
66" – 68"	1.360	1.380	1.400	1.420	1.450	1.470	1.490	1.510	1.530
70" – 72"	1.400	1.420	1.450	1.470	1.490	1.510	1.530	1.550	1.580

GALVANIZED STEEL RECTANGULAR 90 DEGREE ELBOW

Hours per Elbow

One Side Duct Size	One Side Duct Size									
	4"	6"	8"	10"	12"	14"	16"	18"	20"
4"	0.042	0.058	0.078	0.104	0.132	0.212	0.260	0.312	0.370
6"	0.058	0.078	0.104	0.132	0.148	0.232	0.310	0.350	0.390
8"	0.078	0.104	0.132	0.148	0.170	0.260	0.340	0.390	0.415
10"	0.104	0.132	0.148	0.170	0.194	0.292	0.380	0.427	0.448
12"	0.132	0.148	0.170	0.194	0.224	0.328	0.420	0.470	0.487
14" – 16"	0.236	0.266	0.300	0.336	0.340	0.415	0.480	0.519	0.553
18" – 20"	0.339	0.370	0.403	0.438	0.478	0.527	0.555	0.604	0.658
22" – 24"	0.468	0.506	0.549	0.587	0.628	0.867	0.707	0.747	0.786
26" – 28"	0.615	0.655	0.694	0.737	0.775	0.816	0.856	0.895	0.931
30" – 32"	0.854	0.905	0.950	0.997	1.040	1.090	1.140	1.180	1.230
34" – 36"	1.190	1.240	1.280	1.330	1.380	1.430	1.480	1.530	1.580
38" – 40"	1.450	1.490	1.540	1.590	1.640	1.690	1.730	1.780	1.830
42" – 44"	1.730	1.780	1.830	1.870	1.920	1.970	2.020	2.070	2.110
46" – 48"	2.040	2.090	2.130	2.180	2.230	2.280	2.320	2.370	2.420
50"	2.280	2.340	2.390	2.430	2.480	2.530	2.580	2.630	2.670

GALVANIZED STEEL RECTANGULAR 90 DEGREE ELBOW (cont.)

Hours per Elbow

One Side Duct Size	One Side Duct Size									
	22"	24"	26"	28"	30"	32"	34"	36"	38"
22" – 24"	0.828	0.922	1.010	1.100	1.190	1.530	1.670	1.830	1.980
26" – 28"	1.010	1.100	1.190	1.280	1.370	1.730	1.910	2.040	2.180
30" – 32"	1.310	1.410	1.500	1.600	1.700	1.930	2.120	2.260	2.380
34" – 36"	1.710	1.810	1.920	2.030	2.140	2.250	2.360	2.470	2.580
38" – 40"	2.010	2.100	2.210	2.310	2.410	2.510	2.600	2.700	2.810
42" – 44"	2.310	2.410	2.510	2.610	2.710	2.810	2.910	3.010	3.110
46" – 48"	2.610	2.710	2.810	2.910	3.010	3.110	3.210	3.310	3.410
50" – 52"	2.910	3.010	3.110	3.210	3.310	3.410	3.510	3.610	3.710
54" – 56"	3.510	3.620	3.730	3.840	3.950	4.060	4.170	4.270	4.380
58" – 60"	4.260	4.380	4.500	4.610	4.740	4.850	4.970	5.100	5.210
62" – 64"	4.760	4.870	5.020	5.150	5.250	5.380	5.490	5.610	5.740
66" – 68"	5.250	5.380	5.530	5.660	5.790	5.890	6.040	6.150	6.280
70" – 72"	5.760	5.870	6.040	6.150	6.280	6.400	6.510	6.640	6.750

GALVANIZED STEEL RECTANGULAR 90 DEGREE ELBOW *(cont.)*

Hours per Elbow

One Side Duct Size	One Side Duct Size								
	40"	42"	44"	46"	48"	50"	52"	54"	56"
40" – 42"	3.060	3.210	3.360	3.520	3.670	3.820	3.980	4.160	5.100
44" – 46"	3.360	3.520	3.690	3.860	4.010	4.160	4.320	4.460	5.490
48" – 50"	3.660	3.830	4.010	4.160	4.320	4.460	4.610	4.780	5.850
52" – 54"	3.980	4.160	4.320	4.320	4.610	4.780	4.950	5.150	6.260
56" – 58"	5.170	5.340	5.510	5.720	5.910	6.110	6.320	6.510	6.680
60" – 62"	5.590	5.810	6.000	6.190	6.400	6.600	6.790	7.000	7.150
64" – 66"	6.110	6.320	6.510	6.700	6.900	7.090	7.300	7.490	7.660
68" – 70"	6.660	6.850	7.050	7.260	7.410	7.620	7.810	8.010	8.180
72"	7.050	7.220	7.390	7.600	7.770	7.940	8.160	8.410	8.580

GALVANIZED STEEL RECTANGULAR 90 DEGREE ELBOW (cont.)

Hours per Elbow

One Side Duct Size	One Side Duct Size							
	58"	60"	62"	64"	66"	68"	70"	72"
58" – 60"	7.150	7.370	7.620	7.880	8.130	8.410	8.690	8.950
62" – 64"	7.620	7.880	8.130	8.410	8.690	8.950	9.180	9.440
66" – 68"	8.130	8.410	8.690	8.950	9.180	9.440	9.690	9.950
70" – 72"	8.690	8.950	9.180	9.440	9.690	9.950	10.200	10.500

GALVANIZED STEEL RECTANGULAR DROPS AND TAP-IN TEE

Hours per Tee

One Side Duct Size	One Side Duct Size										
	6"	8"	10"	12"	16"	20"	24"	28"	32"
6"	0.072	0.084	0.097	0.109	0.170	0.200	0.231	0.262	0.356
8"	0.084	0.097	0.109	0.121	0.184	0.216	0.247	0.278	0.374
10"	0.097	0.109	0.121	0.133	0.200	0.231	0.262	0.292	0.393
12" – 16"	0.139	0.152	0.167	0.180	0.231	0.262	0.293	0.324	0.432
20" – 24"	0.216	0.231	0.246	0.262	0.293	0.324	0.354	0.385	0.506
28" – 32"	0.309	0.326	0.343	0.360	0.396	0.428	0.462	0.497	0.581
36" – 40"	0.412	0.432	0.450	0.470	0.506	0.544	0.581	0.617	0.655
44" – 48"	0.489	0.506	0.525	0.544	0.581	0.617	0.655	0.694	0.730
52"– 56"	0.591	0.634	0.653	0.675	0.709	0.758	0.798	0.837	0.877
60"	0.747	0.760	0.773	0.794	0.841	0.884	0.927	0.969	1.020

GALVANIZED STEEL SPIRAL DUCTWORK

Hours per Foot

Item	26 Gauge	24 Gauge	22 Gauge	20 Gauge	18 Gauge
3"	0.030	0.041	0.049	0.058	N/A
4"	0.041	0.054	0.065	0.078	N/A
5"	0.051	0.067	0.082	0.097	N/A
6"	0.061	0.081	0.099	0.116	0.155
7"	0.072	0.095	0.115	0.136	0.181
8"	0.082	0.108	0.132	0.154	0.207
9"	0.092	0.122	0.148	0.175	0.233
10"	0.103	0.135	0.165	0.194	0.259
12"	0.123	0.163	0.198	0.233	0.310
14" - 16"	0.174	0.203	0.247	0.291	0.387
18" - 20"	0.195	0.257	0.313	0.369	0.491
22" - 24"	0.236	0.312	0.380	0.446	0.608
26" - 28"	0.277	0.366	0.444	0.525	0.696
30" - 32"	0.318	0.419	0.491	0.604	0.803
34" - 36"	N/A	0.474	0.576	0.681	0.905

GALVANIZED STEEL SPIRAL ELBOW		
Hours per Elbow		
Diameter	90 Degree Elbows	45 Degree Elbows
3"	0.052	0.040
4"	0.088	0.052
5"	0.132	0.076
6"	0.172	0.100
7"	0.232	0.132
8"	0.292	0.172
9"	0.352	0.212
10"	0.470	0.300
12"	0.653	0.400
14" – 16"	1.010	0.576
18" – 20"	1.520	0.850
22" – 24"	2.110	1.190
26" – 28"	2.910	1.610
30" – 32"	3.730	2.060
34" – 36"	5.470	3.010

GALVANIZED STEEL SPIRAL COUPLING

Hours per Coupling

Diameter	Hours
3"	0.020
4"	0.024
5"	0.028
6"	0.036
7"	0.040
8"	0.048
9"	0.104
10"	0.116
12"	0.140
14" – 16"	0.174
18" – 20"	0.220
22" – 24"	0.266
26" – 28"	0.312
30" – 32"	0.357
34" – 36"	0.403

GALVANIZED STEEL SPIRAL REDUCER	
Hours per Reducer	
Diameter	Hours
3"	0.040
4"	0.048
5"	0.056
6"	0.072
7"	0.080
8"	0.096
9"	0.132
10"	0.176
12"	0.212
14" – 16"	0.262
18" – 20"	0.330
22" – 24"	0.398
26" – 28"	0.465
30" – 32"	0.536
34" – 36"	0.606

GALVANIZED STEEL SPIRAL TEE WITH REDUCING BRANCH

Hours per Tee

Diameter	Reducing Branch Dimension															
	3"	4"	5"	6"	7"	8"	9"	10"	12"	14"	16"	18"	20"	22"
4"	0.109	N/A	N/A	N/A	N/A	N/A	N/A	N/A	N/A	N/A	N/A	N/A	N/A	N/A
5"	0.112	0.120	N/A	N/A	N/A	N/A	N/A	N/A	N/A	N/A	N/A	N/A	N/A	N/A
6"	0.128	0.140	0.148	N/A	N/A	N/A	N/A	N/A	N/A	N/A	N/A	N/A	N/A	N/A
7"	0.148	0.156	0.164	0.180	N/A	N/A	N/A	N/A	N/A	N/A	N/A	N/A	N/A	N/A
8"	0.164	0.176	0.188	0.200	0.164	N/A	N/A	N/A	N/A	N/A	N/A	N/A	N/A	N/A
9"	0.180	0.192	0.204	0.220	0.180	0.228	N/A	N/A	N/A	N/A	N/A	N/A	N/A	N/A
10"	0.256	0.268	0.284	0.300	0.252	0.268	0.284	N/A	N/A	N/A	N/A	N/A	N/A	N/A
12"	0.308	0.324	0.340	0.356	0.304	0.324	0.340	0.360	N/A	N/A	N/A	N/A	N/A	N/A
14" – 16"	0.382	0.403	0.428	0.449	0.381	0.401	0.426	0.447	0.572	0.572	N/A	N/A	N/A	N/A
18" – 20"	0.485	0.508	0.538	0.566	0.482	0.508	0.540	0.564	0.681	0.681	0.739	0.837	N/A	N/A
22" – 24"	0.585	0.619	0.651	0.683	0.579	0.617	0.649	0.683	0.820	0.820	0.888	0.954	1.020	1.570
26" – 28"	0.683	0.724	0.764	0.807	0.683	0.722	0.760	0.805	0.965	0.965	1.040	1.120	1.200	1.320
30" – 32"	0.786	0.830	0.873	0.924	0.784	0.830	0.875	0.924	1.110	1.110	1.200	1.290	1.380	1.480
34" – 36"	1.020	1.080	1.150	1.200	1.020	1.080	1.150	1.200	1.440	1.440	1.560	1.680	1.810	1.920

GALVANIZED STEEL SPIRAL TEE WITH REDUCING RUN AND BRANCH

Hours per Tee

Largest Run Diameter	Branch Diameter													
	3"	4"	5"	6"	7"	8"	9"	10"	12"	14"	16"	18"	20"	22"
4"	0.109	N/A	N/A	N/A	N/A	N/A	N/A	N/A	N/A	N/A	N/A	N/A	N/A	N/A
5"	0.112	0.120	N/A	N/A	N/A	N/A	N/A	N/A	N/A	N/A	N/A	N/A	N/A	N/A
6"	0.128	0.140	0.148	N/A	N/A	N/A	N/A	N/A	N/A	N/A	N/A	N/A	N/A	N/A
7"	0.148	0.156	0.164	0.180	N/A	N/A	N/A	N/A	N/A	N/A	N/A	N/A	N/A	N/A
8"	0.164	0.176	0.188	0.200	0.212	N/A	N/A	N/A	N/A	N/A	N/A	N/A	N/A	N/A
9"	0.180	0.192	0.204	0.220	0.232	0.248	N/A	N/A	N/A	N/A	N/A	N/A	N/A	N/A
10"	0.256	0.268	0.284	0.300	0.312	0.328	0.344	N/A	N/A	N/A	N/A	N/A	N/A	N/A
12"	0.308	0.324	0.340	0.356	0.376	0.396	0.412	0.427	N/A	N/A	N/A	N/A	N/A	N/A
14" – 16"	0.382	0.403	0.428	0.449	0.470	0.493	0.515	0.538	0.581	0.666	N/A	N/A	N/A	N/A
18" – 20"	0.485	0.508	0.538	0.566	0.636	0.623	0.653	0.681	0.739	0.794	0.833	0.956	N/A	N/A
22" – 24"	0.585	0.619	0.651	0.683	0.717	0.751	0.779	0.820	0.888	0.959	1.030	1.090	1.160	1.280
26" – 28"	0.683	0.724	0.764	0.807	0.845	0.882	0.924	0.965	1.040	1.110	1.200	1.290	1.370	1.450
30" – 32"	0.786	0.830	0.873	0.924	0.969	1.020	1.060	1.110	1.200	1.290	1.390	1.480	1.570	1.640
34" – 36"	1.020	1.080	1.150	1.200	1.270	1.330	1.390	1.440	1.570	1.690	1.810	1.930	2.030	2.160

GALVANIZED STEEL SPIRAL CROSS WITH REDUCING BRANCH

Hours per Cross

Item	Reducing Branch												
	3"	4"	5"	6"	7"	8"	9"	12"	14"	16"	18"	20"	22"
4"	0.104	N/A	N/A	N/A	N/A	N/A	N/A	N/A	N/A	N/A	N/A	N/A	N/A
5"	0.128	0.144	N/A	N/A	N/A	N/A	N/A	N/A	N/A	N/A	N/A	N/A	N/A
6"	0.152	0.168	0.184	N/A	N/A	N/A	N/A	N/A	N/A	N/A	N/A	N/A	N/A
7"	0.176	0.196	0.216	0.232	N/A	N/A	N/A	N/A	N/A	N/A	N/A	N/A	N/A
8"	0.196	0.216	0.240	0.260	0.280	N/A	N/A	N/A	N/A	N/A	N/A	N/A	N/A
9"	0.216	0.240	0.264	0.288	0.308	0.310	N/A	N/A	N/A	N/A	N/A	N/A	N/A
10"	0.304	0.328	0.356	0.380	0.408	0.403	0.427	N/A	N/A	N/A	N/A	N/A	N/A
12"	0.352	0.384	0.412	0.440	0.470	0.470	0.495	N/A	N/A	N/A	N/A	N/A	N/A
14" – 16"	0.428	0.461	0.497	0.532	0.566	0.561	0.596	0.692	1.810	N/A	N/A	N/A	N/A
18" – 20"	0.525	0.568	0.613	0.653	0.692	0.690	0.732	0.850	2.110	1.080	1.220	N/A	N/A
22" – 24"	0.634	0.685	0.741	0.790	0.841	0.835	0.882	1.030	2.550	1.300	1.400	1.510	1.680
26" – 28"	0.745	0.805	0.867	0.927	0.986	0.980	1.040	1.180	2.980	1.530	1.650	1.770	1.890
30" – 32"	0.856	0.927	0.993	1.060	1.130	1.120	1.190	1.380	3.430	1.760	1.890	2.050	2.170
34" – 36"	1.110	1.200	1.290	1.380	1.480	1.470	1.550	1.800	4.470	2.310	2.470	2.650	2.840

DUCT INSULATION

Fiberglass (Hours per Sq. Ft)

Duct	Duct Liner		
	1", 1.5 lbs	1", 3 lbs	2", 1.5 lbs
26 Gauge, Size 0-12	0.042	0.048	0.048
24 Gauge, Size 13-30	0.033	0.038	0.038
22 Gauge, Size 31-54	0.029	0.033	0.033
20 Gauge, Size 55-84	0.024	0.028	0.028
18 Gauge, Size 85-up	0.021	0.024	0.024

Rigid Board (Hours per Sq. Ft)

Duct	Duct Liner		
	1"	1½"	2"
26 Gauge, Size 0-12	0.063	0.073	0.084
24 Gauge, Size 13-30	0.050	0.058	0.066
22 Gauge, Size 31-54	0.043	0.051	0.058
20 Gauge, Size 55-84	0.036	0.042	0.048
18 Gauge, Size 85-up	0.032	0.037	0.042

Blanket (Hours per Sq. Ft)

Duct	Duct Liner			
	½"	1"	1½"	2"
26 Gauge, Size 0-12	0.036	0.036	0.039	0.039
24 Gauge, Size 13-30	0.028	0.028	0.031	0.031
22 Gauge, Size 31-54	0.025	0.025	0.028	0.028
20 Gauge, Size 55-84	0.020	0.020	0.023	0.023
18 Gauge, Size 85-up	0.018	0.018	0.021	0.021

DUCTWORK ACCESSORIES

Damper for Rectangular Ductwork

Size	Hours per Damper
12" × 6"	0.60
12" × 10"	0.65
12" × 12"	0.70
18" × 12"	0.80
18" × 18"	0.85
24" × 12"	0.85
24" × 18"	0.90
24" × 24"	1.00
30" × 12"	0.90
30" × 18"	1.00
30" × 24"	1.10
30" × 30"	1.20
36" × 18"	1.10
36" × 36"	1.40
42" × 24"	1.30
42" × 42"	1.50
48" × 24"	1.40
48" × 48"	1.60
54" × 24"	1.45
60" × 24"	1.70

Damper for Round Ductwork

Size	Hours per Damper
6"	0.60
8"	0.55
10"	0.60
12"	0.65
14"	0.70
16"	0.77
18"	0.84
20"	0.90

DUCTWORK ACCESSORIES (cont.)

Fire Damper

Size	Curtain Type	UL Sleeve
12" × 6"	1.00	1.20
12" × 10"	1.05	1.26
12" × 12"	1.10	1.32
18" × 12"	1.30	1.56
18" × 18"	1.30	1.56
24" × 12"	1.40	1.68
24" × 18"	1.60	1.92
24" × 24"	1.80	2.16
30" × 12"	1.50	1.80
30" × 18"	1.75	2.10
30" × 24"	2.00	2.40
30" × 30"	2.00	2.40
36" × 18"	2.00	2.40
36" × 36"	2.30	2.64
42" × 24"	2.40	2.88
42" × 42"	2.50	3.00
48" × 24"	2.60	3.12
48" × 48"	3.50	4.20
54" × 24"	2.80	3.36
60" × 24"	3.00	3.60

DUCTWORK ACCESSORIES *(cont.)*	
Diffuser	
Size	**Hours per Diffuser**
6", 250 CFM	0.35
8", 400 CFM	0.35
10", 600 CFM	0.35
12", 750 CFM	0.35
14", 1,100 CFM	0.35
16", 1,250 CFM	0.45
18", 1,550 CFM	0.45
20", 2,100 CFM	0.55
24", 2,500 CFM	0.65
Grille	
Size	**Hours per Grille**
12" × 6"	0.80
12" × 12"	0.90
18" × 12"	1.00
18" × 18"	1.10
24" × 12"	1.10
24" × 24"	1.30
30" × 12"	1.20
30" × 24"	1.50
36" × 18"	1.50
42" × 24"	1.60
48" × 24"	1.70
48" × 48"	2.60
54" × 24"	2.00
60" × 24"	2.30

DUCTWORK ACCESSORIES (cont.)

Belt Guard – Galvanized (Hours per Pound)

Nominal Wheel Diameter	Fabrication	Erection	Fabrication & Erection
12"	2.00	0.50	2.50
15"	2.00	0.50	2.50
16"	2.00	0.50	2.50
18"	2.15	0.65	2.80
20"	2.30	0.75	3.05
22"	2.45	0.75	3.20
24"	2.60	0.90	3.50
28"	2.75	1.00	3.75
30"	2.90	1.00	3.90
34"	3.05	1.00	4.05
36"	3.20	1.15	4.35
40"	3.75	1.40	5.15
44"	4.30	1.65	5.95
48"	4.90	2.00	6.90
54"	5.50	2.30	7.80
60"	6.00	2.65	8.65
66"	6.50	3.00	9.50
72"	7.00	3.50	10.50
80"	8.00	4.00	12.00

Belt Guard – Flex (Hours per Item)

Length	Fabrication	Erection	Fabrication & Erection
3'	3.00	2.00	5.00
5'	4.00	3.00	7.00
7'	5.00	4.00	9.00
10'	6.00	5.00	11.00

DUCTWORK ACCESSORIES *(cont.)*

Spin-in Galvanized Steel Collar

Size	Hours per Collar	
	Without Damper	**With Damper**
4"	0.20	0.25
5"	0.20	0.25
6"	0.20	0.25
7"	0.20	0.25
8"	0.25	0.30
9"	0.25	0.30
10"	0.30	0.35
12"	0.35	0.40
14"	0.40	0.45
16"	0.45	0.50
18"	0.50	0.55

DUCTWORK ACCESSORIES (cont.)

Terminal Boxes

Size	Fan Powered		VAV/Mixing/Induction	
	No Reheat Coil	With Reheat Coil	No Reheat Coil	With Reheat Coil
200-400 CFM	2.3	2.8	2.0	2.7
400-600 CFM	3.1	3.4	2.7	2.9
600-800 CFM	3.8	4.1	3.3	3.6
800-1,000 CFM	4.6	4.9	4.0	4.3
1,000-1,500 CFM	5.2	5.5	4.5	4.9
1,500-2,000 CFM	5.8	6.2	5.0	5.5
2,000-3,000 CFM	6.3	6.7	5.5	6.0

DUCTWORK ACCESSORIES *(cont.)*		
Louvers - Rectangular		
Size	**Aluminum**	**Galvanized**
12" × 6"	1.10	1.32
12" × 12"	1.20	1.44
18" × 12"	1.30	1.56
18" × 18"	1.50	1.80
24" × 12"	1.50	1.80
30" × 12"	2.00	2.40
30" × 24"	2.30	2.76
36" × 18"	2.30	2.76
42" × 24"	2.60	3.12
48" × 24"	2.80	3.36
48" × 48"	4.00	4.80
54" × 24"	3.30	3.96
60" × 24"	3.50	4.20
72" × 48"	5.00	6.00
84" × 36"	5.00	6.00
84" × 84"	7.00	8.40
96" × 48"	6.00	7.20
96" × 98"	7.50	9.00

DUCTWORK ACCESSORIES *(cont.)*		
Louvers - Round		
Size	**Surface Mount**	**Lay-in**
6"	0.70	0.40
8"	0.80	0.50
10"	0.90	0.50
12"	1.00	0.60
14"	1.10	0.70
16"	1.20	0.70
18"	1.30	0.80
20"	1.50	0.90
24"	1.60	1.00
30"	1.90	1.10
36"	2.20	1.30
Access Doors		
Size	**Hours per Door**	
8" × 8"	0.40	
12" × 12"	0.65	
12" × 16"	0.70	
16" × 16"	0.75	
16" × 24"	1.00	
18" × 36"	1.20	
22" × 36"	1.35	
22" × 46"	1.55	
22" × 58"	1.68	

DUCTWORK ACCESSORIES (cont.)

Duct to Equipment Flex Connection

Size	Hours per Connection
18" × 12"	0.70
18" × 18"	0.85
24" × 12"	0.85
24" × 18"	1.00
24" × 24"	1.30
30" × 12"	1.00
30" × 18"	1.30
30" × 30"	1.60
36" × 12"	1.30
36" × 24"	1.60
36" × 36"	1.80
48" × 24"	1.80
48" × 48"	2.00
60" × 42"	2.00
84" × 60"	3.20
108" × 72"	4.10

Apparatus Housing (Hours per Sq. Ft)

Item	Fabrication	Erection	Fabrication & Erection
Single-skin panels	0.04	0.25	0.29
Double-skin panels	0.09	0.30	0.39
Add for access doors	0.40	0.24	0.64

COPPER PIPE	
Item/Size	Hours per Foot
Type K with brazed joints	
1/2"	0.12
3/4"	0.13
1"	0.15
1 1/4"	0.17
1 1/2"	0.19
2"	0.22
2 1/2"	0.25
3"	0.28
4"	0.32
Type K with soft-soldered joints	
1/2"	0.11
3/4"	0.12
1"	0.14
1 1/4"	0.16
1 1/2"	0.18
2"	0.21
2 1/2"	0.24
3"	0.27
4"	0.31
Type L with brazed joints	
1/2"	0.12
3/4"	0.13
1"	0.15
1 1/4"	0.17
1 1/2"	0.19
2"	0.22
2 1/2"	0.25
3"	0.28
4"	0.32

COPPER PIPE *(cont.)*	
Item/Size	**Hours per Foot**
Type L with soft-soldered joints	
1/2"	0.11
3/4"	0.12
1"	0.14
1 1/4"	0.16
1 1/2"	0.18
2"	0.21
2 1/2"	0.24
3"	0.27
4"	0.31
Type M with brazed joints	
1/2"	0.11
3/4"	0.12
1"	0.14
1 1/4"	0.16
1 1/2"	0.18
2"	0.21
2 1/2"	0.24
3"	0.27
4"	0.31
Type M with soft-soldered joints	
1/2"	0.11
3/4"	0.12
1"	0.14
1 1/4"	0.16
1 1/2"	0.18
2"	0.21
2 1/2"	0.24
3"	0.27
4"	0.31

PLASTIC PIPE	
Item/Size	**Hour per Foot**
Schedule 40	
1/2"	0.08
3/4"	0.09
1"	0.10
1 1/4"	0.11
1 1/2"	0.12
2"	0.13
2 1/2"	0.13
3"	0.15
4"	0.18
6"	0.23
8"	0.27
Schedule 80	
1/2"	0.09
3/4"	0.09
1"	0.10
1 1/4"	0.12
1 1/2"	0.12
2"	0.12
2 1/2"	0.13
3"	0.13
4"	0.16
6"	0.25
8"	0.30

STEEL PIPE	
Item/Size	Hours per Foot
Schedule 40, threaded joints	
1/2"	0.12
3/4"	0.13
1"	0.14
1 1/4"	0.16
1 1/2"	0.18
2"	0.20
2 1/2"	0.24
3"	0.27
4"	0.40
Schedule 80, threaded joints	
1/2"	0.12
3/4"	0.13
1"	0.15
1 1/4"	0.16
1 1/2"	0.19
2"	0.21
2 1/2"	0.30
3"	0.34
4"	0.41

STEEL PIPE *(cont.)*	
Item/Size	Hours per Foot
Schedule 10, grooved joints	
2"	0.18
2½"	0.19
3"	0.22
4"	0.26
6"	0.34
8"	0.40
10"	0.52
12"	0.60
Schedule 40, grooved joints	
2"	0.18
2½"	0.19
3"	0.22
4"	0.26
6"	0.35
8"	0.41
10"	0.53
12"	0.62
Schedule 80, grooved joints	
2"	0.20
2½"	0.21
3"	0.24
4"	0.28
6"	0.38
8"	0.45
10"	0.58
12"	0.67

PIPING ACCESSORIES	
Valve	
Item/Size	Hours per Valve
Bronze, brazed joints	
1/2"	0.24
3/4"	0.30
1"	0.36
1 1/4"	0.48
1 1/2"	0.54
2"	0.60
2 1/2"	1.00
3"	1.50
Bronze, soft-soldered joints	
1/2"	0.20
3/4"	0.25
1"	0.30
1 1/4"	0.40
1 1/2"	0.45
2"	0.50
2 1/2"	0.83
3"	1.24

PIPING ACCESSORIES *(cont.)*	
Valve	
Item/Size	**Hours per Valve**
Bronze, threaded joints	
½"	0.21
¾"	0.25
1"	0.30
1¼"	0.40
1½"	0.45
2"	0.50
2½"	0.75
3"	0.95
Iron, flanged joints	
2"	0.50
2½"	0.60
3"	0.75
4"	1.35
5"	2.00
6"	2.50
8"	3.00
10"	4.00
12"	4.50

PIPING ACCESSORIES *(cont.)*	
2-Way Control Valve	
Item/Size	**Hours per Valve**
Bronze, threaded joints	
1/2 "	0.21
3/4 "	0.28
1"	0.35
1 1/4 "	0.43
1 1/2 "	0.51
2"	0.68
2 1/2 "	0.83
3"	0.99
4"	1.35
6"	2.50
8"	3.00
10"	4.00
12"	4.50
Iron, flanged joints	
2"	0.50
2 1/2 "	0.60
3"	0.75
4"	1.35
6"	2.50
8"	3.00
10"	4.00
12"	4.50

PIPING ACCESSORIES (cont.)	
3-Way Control Valve	
Item/Size	Hours per Valve
Bronze, threaded joints	
1/2"	0.26
3/4"	0.37
1"	0.48
1 1/4"	0.58
1 1/2"	0.68
2"	0.91
2 1/2"	1.12
3"	1.33
4"	2.00
6"	3.70
8"	4.40
10"	5.90
12"	6.50
Iron, flanged joints	
2"	0.91
2 1/2"	1.12
3"	1.33
4"	2.00
6"	3.70
8"	4.40
10"	5.90
12"	6.50

PIPING ACCESSORIES *(cont.)*	
Companion Flanges	
Item/Size	**Hours per Flange**
Steel, welded joint	
2½"	0.81
3"	0.98
4"	1.30
6"	1.95
8"	2.35
10"	2.90
12"	3.50
Steel, threaded joint	
2"	0.29
2½"	0.38
3"	0.46
4"	0.60
6"	0.68
8"	0.76

PIPING ACCESSORIES *(cont.)*	
Companion Flanges *(cont.)*	
Item/Size	**Hours per Flange**
PVC	
2"	0.14
2½"	0.18
3"	0.21
4"	0.28
5"	0.31
6"	0.42
8"	0.56
Bolt and Gasket Sets	
Item/Size	**Hours per Set**
2"	0.50
2½"	0.65
3"	0.75
4"	1.00
5"	1.10
6"	1.20
8"	1.25
10"	1.70
12"	2.20

PIPING ACCESSORIES *(cont.)*	
Hanger	
Item	**Hours per Hanger**
For copper pipes	
1/2"	0.25
3/4"	0.25
1"	0.25
1 1/4"	0.30
1 1/2"	0.30
2"	0.30
2 1/2"	0.35
3"	0.35
4"	0.35
For plastic pipes	
1/2"	0.25
3/4"	0.25
1"	0.25
1 1/4"	0.30
1 1/2"	0.30
2"	0.30
2 1/2"	0.35
3"	0.35
4"	0.35
5"	0.45
6"	0.45
8"	0.45

PIPING ACCESSORIES *(cont.)*	
Hanger *(cont.)*	
Item	**Hours per Hanger**
For steel pipes	
$\frac{1}{2}$"	0.25
$\frac{3}{4}$"	0.25
1"	0.25
$1\frac{1}{4}$"	0.30
$1\frac{1}{2}$"	0.30
2"	0.30
$2\frac{1}{2}$"	0.35
3"	0.35
4"	0.35
5"	0.45
6"	0.45
8"	0.45
10"	0.55
12"	0.60

PIPING ACCESSORIES *(cont.)*	
Riser Clamp	
Item	Hours per Clamp
For copper pipes	
1/2"	0.10
3/4"	0.10
1"	0.10
1 1/4"	0.11
1 1/2"	0.11
2"	0.12
2 1/2"	0.12
3"	0.12
4"	0.13
For plastic pipes	
1/2"	0.25
3/4"	0.25
1"	0.25
1 1/4"	0.30
1 1/2"	0.30
2"	0.30
2 1/2"	0.35
3"	0.35
4"	0.35
5"	0.45
6"	0.45
8"	0.45

PIPING ACCESSORIES *(cont.)*	
Riser Clamp *(cont.)*	
Item	**Hours per Clamp**
For steel pipes	
$1/2$"	0.10
$3/4$"	0.10
1"	0.10
$1\frac{1}{4}$"	0.11
$1\frac{1}{2}$"	0.11
2"	0.12
$2\frac{1}{2}$"	0.12
3"	0.12
4"	0.13
5"	0.18
6"	0.20
8"	0.20
10"	0.25
12"	0.25

PIPING ACCESSORIES *(cont.)*

Pipe Insulation

Pipe Size	Hours per Linear Foot
1/2"	0.040
3/4"	0.040
1"	0.042
1 1/4"	0.042
1 1/2"	0.044
2"	0.046
2 1/2"	0.050
3"	0.056
4"	0.069
5"	0.072
6"	0.085
8"	0.129
10"	0.154
12"	0.171

PIPING ACCESSORIES *(cont.)*	
Pipe Painting	
Item/Size	**Hours per Linear Foot**
½"	0.19
¾"	0.20
1"	0.21
1¼"	0.22
1½"	0.22
2"	0.23
2½"	0.24
3"	0.26
4"	0.28
5"	0.32
6"	0.35
8"	0.38
10"	0.40
12"	0.42

AIR BALANCING	
Air Balancing — Terminal Boxes	
Item	Hours per Item
Variable air volume	0.40
Constant volume	0.40
Dual duct box	0.70
Terminal box and all downstream diffusers	1.75
Air Balancing — Diffusers	
Item	Hours per Item
Ceiling diffusers to 500 CFM	0.25
Ceiling diffusers over 500 CFM	0.35
Ceiling diffusers perforated	0.37
Sidewall diffusers double deflection	0.25
Linear slot diffusers 2 slot	0.06
Linear slot diffusers 3 slot	0.07
Linear slot diffusers 4 slot	0.08
Plug-in diffuser	0.25
Light troffers 1 slot	0.20
Light troffers 2 slot	0.22
Laminar flow diffusers	0.75
Laminar flow diffusers with HEPA filter	3.00

AIR BALANCING *(cont.)*	
Air Balancing — Grilles	
Item	**Hours per Item**
Return air grilles to 800 CFM	0.25
Return air grilles over 800 CFM	0.35
Bar type return air grilles 2 slot	0.06
Bar type return air grilles 4 slot	0.07
Bar type return air grilles 6 slot	0.08
Air Balancing — Centrifugal Fans	
Item	**Hours per Item**
Supply and return air to 3,000 CFM	1.00
Supply and return air to 5,000 CFM	1.50
Supply and return air to 10,000 CFM	2.00
Supply and return air to 20,000 CFM	2.75
Roof exhaust to 6,000 CFM	0.73
Roof exhaust to 10,000 CFM	0.93
Roof exhaust to 20,000 CFM	1.50
Range hood exhaust to 10,000 CFM	1.75
Range hood exhaust to 20,000 CFM	2.50

AIR BALANCING (cont.)

Air Balancing — Fan Coil Units

Item	Hours per Item
Vertical floor model	0.50
Horizontal ceiling mounted	0.65

Air Balancing — Fume Hoods

Item	Hours per Item
Lab fume hood	2.00
General purpose	0.45
Dust collectors	1.50

AIR BALANCING *(cont.)*	
Wet Balancing	
Item	**Hours per Item**
Boiler	1.65
Chiller	3.50
Cooling tower	2.00
Circulating pump	1.65
Heat exchanger	1.75
AHU water coil	2.50
Reheat coil	0.45
Wall fin radiation	0.35
Radiant heat panel	0.45
Fan coil unit, 2 pipe	0.35
Fan coil unit, 4 pipe	0.45
Cabinet heater	0.35
Unit ventilator	0.45
Unit heater	0.35

CHAPTER 7
Technical Reference

Disclaimer: Data tables in this chapter represent the author's best judgment and care for the information published. Instructions for these data tables, when given, should be carefully studied before use. The numerical results for any data table is affected by numerous specific project factors such as design standards, site conditions, labor productivity, material waste, etc. Please also refer to the governing mechanical codes for your specific jobs. If anything in this book conflicts with your code, the code should always govern. Neither the author nor the publisher is responsible for any losses or damages with respect to the accuracy, correctness, value, or sufficiency of the data, methods, or other information contained herein.

AREAS OF COMMON GEOMETRIC SHAPES

Shape	Formula
	Parallelogram Area = a × b
	Trapezoid Area = c × $\frac{1}{2}$ × (a + b)
 hypotenuse (c) = $\sqrt{(a^2 + b^2)}$	**Right Triangle** Area = $\frac{1}{2}$ × a × b
	Regular Triangle Area = $\frac{1}{2}$ × a × b
 b (Diameter) = 2 × a (Radius)	**Circle** Area = 3.1416 × a^2 Circumference = 3.1416 × b = 6.2832 × a

7-2

VOLUMES OF COMMON GEOMETRIC SHAPES

Shape	Formula
	Cylinder Volume = $3.1416 \times a/2 \times a/2 \times b$ $= 0.7854 \times a^2 \times b$
	Pyramid Volume = $\frac{1}{3} \times a \times b \times c$
	Cone Volume = $\frac{1}{3} \times 3.1416 \times b \times b \times c$ $= 1.0472 \times b^2 \times c$ Or Volume = $0.3518 \times a^2 \times c$
	Sphere Volume = $\frac{1}{6} \times 3.1416 \times a \times a \times a$ $= 0.5236 \times a^3$

SQUARE, CUBE, SQUARE ROOT, AND CUBIC ROOT FOR NUMBERS FROM 1 TO 100

Number	Square	Cube	Square Root	Cubic Root
1	1	1	1.0000	1.0000
2	4	8	1.4142	1.2599
3	9	27	1.7321	1.4422
4	16	64	2.0000	1.5874
5	25	125	2.2361	1.7100
6	36	216	2.4495	1.8171
7	49	343	2.6458	1.9129
8	64	512	2.8284	2.0000
9	81	729	3.0000	2.0801
10	100	1000	3.1623	2.1544
11	121	1331	3.3166	2.2240
12	144	1728	3.4641	2.2894
13	169	2197	3.6056	2.3513
14	196	2744	3.7417	2.4101
15	225	3375	3.8730	2.4662
16	256	4096	4.0000	2.5198
17	289	4913	4.1231	2.5713
18	324	5832	4.2426	2.6207
19	361	6859	4.3589	2.6684
20	400	8000	4.4721	2.7144
21	441	9261	4.5826	2.7589
22	484	10648	4.6904	2.8020
23	529	12167	4.7958	2.8439
24	576	13824	4.8990	2.8845
25	625	15625	5.0000	2.9240

Number	Square	Cube	Square Root	Cubic Root
26	676	17576	5.0990	2.9625
27	729	19683	5.1962	3.0000
28	784	21952	5.2915	3.0366
29	841	24389	5.3852	3.0723
30	900	27000	5.4772	3.1072
31	961	29791	5.5678	3.1414
32	1024	32768	5.6569	3.1748
33	1089	35937	5.7446	3.2075
34	1156	39304	5.8310	3.2396
35	1225	42875	5.9161	3.2711
36	1296	46656	6.0000	3.3019
37	1369	50653	6.0828	3.3322
38	1444	54872	6.1644	3.3620
39	1521	59319	6.2450	3.3912
40	1600	64000	6.3246	3.4200
41	1681	68921	6.4031	3.4482
42	1764	74088	6.4807	3.4760
43	1849	79507	6.5574	3.5034
44	1936	85184	6.6332	3.5303
45	2025	91125	6.7082	3.5569
46	2116	97336	6.7823	3.5830
47	2209	103823	6.8557	3.6088
48	2304	110592	6.9282	3.6342
49	2401	117649	7.0000	3.6593
50	2500	125000	7.0711	3.6840

Number	Square	Cube	Square Root	Cubic Root
51	2601	132651	7.1414	3.7084
52	2704	140608	7.2111	3.7325
53	2809	148877	7.2801	3.7563
54	2916	157464	7.3485	3.7798
55	3025	166375	7.4162	3.8030
56	3136	175616	7.4833	3.8259
57	3249	185193	7.5498	3.8485
58	3364	195112	7.6158	3.8709
59	3481	205379	7.6811	3.8930
60	3600	216000	7.7460	3.9149
61	3721	226981	7.8102	3.9365
62	3844	238328	7.8740	3.9579
63	3969	250047	7.9373	3.9791
64	4096	262144	8.0000	4.0000
65	4225	274625	8.0623	4.0207
66	4356	287496	8.1240	4.0412
67	4489	300763	8.1854	4.0615
68	4624	314432	8.2462	4.0817
69	4761	328509	8.3066	4.1016
70	4900	343000	8.3666	4.1213
71	5041	357911	8.4261	4.1408
72	5184	373248	8.4853	4.1602
73	5329	389017	8.5440	4.1793
74	5476	405224	8.6023	4.1983
75	5625	421875	8.6603	4.2172

SQUARE, CUBE, SQUARE ROOT, AND CUBIC ROOT
FOR NUMBERS FROM 1 TO 100 (*cont.*)

Number	Square	Cube	Square Root	Cubic Root
76	5776	438976	8.7178	4.2358
77	5929	456533	8.7750	4.2543
78	6084	474552	8.8318	4.2727
79	6241	493039	8.8882	4.2908
80	6400	512000	8.9443	4.3089
81	6561	531441	9.0000	4.3267
82	6724	551368	9.0554	4.3445
83	6889	571787	9.1104	4.3621
84	7056	592704	9.1652	4.3795
85	7225	614125	9.2195	4.3968
86	7396	636056	9.2736	4.4140
87	7569	658503	9.3274	4.4310
88	7744	681472	9.3808	4.4480
89	7921	704969	9.4340	4.4647
90	8100	729000	9.4868	4.4814
91	8281	753571	9.5394	4.4979
92	8464	778688	9.5917	4.5144
93	8649	804357	9.6437	4.5307
94	8836	830584	9.6954	4.5468
95	9025	857375	9.7468	4.5629
96	9216	884736	9.7980	4.5789
97	9409	912673	9.8489	4.5947
98	9604	941192	9.8995	4.6104
99	9801	970299	9.9499	4.6261
100	10000	1000000	10.0000	4.6416

CIRCLE CIRCUMFERENCE AND AREA
(DIAMETERS FROM 1 TO 100)

Diameter	Radius	Circumference	Area
(based on the same type of units)			
1	0.5	3.1416	0.7854
2	1.0	6.2832	3.1416
3	1.5	9.4248	7.0686
4	2.0	12.5664	12.5664
5	2.5	15.7080	19.6350
6	3.0	18.8496	28.2744
7	3.5	21.9912	38.4846
8	4.0	25.1328	50.2656
9	4.5	28.2744	63.6174
10	5.0	31.4160	78.5400
11	5.5	34.5576	95.0334
12	6.0	37.6992	113.0976
13	6.5	40.8408	132.7326
14	7.0	43.9824	153.9384
15	7.5	47.1240	176.7150
16	8.0	50.2656	201.0624
17	8.5	53.4072	226.9806
18	9.0	56.5488	254.4696
19	9.5	59.6904	283.5294
20	10.0	62.8320	314.1600
21	10.5	65.9736	346.3614
22	11.0	69.1152	380.1336
23	11.5	72.2568	415.4766
24	12.0	75.3984	452.3904
25	12.5	78.5400	490.8750

CIRCLE CIRCUMFERENCE AND AREA
(DIAMETERS FROM 1 TO 100) (cont.)

Diameter	Radius	Circumference	Area
(based on the same type of units)			
26	13.0	81.6816	530.9304
27	13.5	84.8232	572.5566
28	14.0	87.9648	615.7536
29	14.5	91.1064	660.5214
30	15.0	94.2480	706.8600
31	15.5	97.3896	754.7694
32	16.0	100.5312	804.2496
33	16.5	103.6728	855.3006
34	17.0	106.8144	907.9224
35	17.5	109.9560	962.1150
36	18.0	113.0976	1017.8784
37	18.5	116.2392	1075.2126
38	19.0	119.3808	1134.1176
39	19.5	122.5224	1194.5934
40	20.0	125.6640	1256.6400
41	20.5	128.8056	1320.2574
42	21.0	131.9472	1385.4456
43	21.5	135.0888	1452.2046
44	22.0	138.2304	1520.5344
45	22.5	141.3720	1590.4350
46	23.0	144.5136	1661.9064
47	23.5	147.6552	1734.9486
48	24.0	150.7968	1809.5616
49	24.5	153.9384	1885.7454
50	25.0	157.0800	1963.5000

CIRCLE CIRCUMFERENCE AND AREA
(DIAMETERS FROM 1 TO 100) *(cont.)*

Diameter	Radius	Circumference	Area
(based on the same type of units)			
51	25.5	160.2216	2042.8254
52	26.0	163.3632	2123.7216
53	26.5	166.5048	2206.1886
54	27.0	169.6464	2290.2264
55	27.5	172.7880	2375.8350
56	28.0	175.9296	2463.0144
57	28.5	179.0712	2551.7646
58	29.0	182.2128	2642.0856
59	29.5	185.3544	2733.9774
60	30.0	188.4960	2827.4400
61	30.5	191.6376	2922.4734
62	31.0	194.7792	3019.0776
63	31.5	197.9208	3117.2526
64	32.0	201.0624	3216.9984
65	32.5	204.2040	3318.3150
66	33.0	207.3456	3421.2024
67	33.5	210.4872	3525.6606
68	34.0	213.6288	3631.6896
69	34.5	216.7704	3739.2894
70	35.0	219.9120	3848.4600
71	35.5	223.0536	3959.2014
72	36.0	226.1952	4071.5136
73	36.5	229.3368	4185.3966
74	37.0	232.4784	4300.8504
75	37.5	235.6200	4417.8750

CIRCLE CIRCUMFERENCE AND AREA
(DIAMETERS FROM 1 TO 100) *(cont.)*

Diameter	Radius	Circumference	Area
(based on the same type of units)			
76	38.0	238.7616	4536.4704
77	38.5	241.9032	4656.6366
78	39.0	245.0448	4778.3736
79	39.5	248.1864	4901.6814
80	40.0	251.3280	5026.5600
81	40.5	254.4696	5153.0094
82	41.0	257.6112	5281.0296
83	41.5	260.7528	5410.6206
84	42.0	263.8944	5541.7824
85	42.5	267.0360	5674.5150
86	43.0	270.1776	5808.8184
87	43.5	273.3192	5944.6926
88	44.0	276.4608	6082.1376
89	44.5	279.6024	6221.1534
90	45.0	282.7440	6361.7400
91	45.5	285.8856	6503.8974
92	46.0	289.0272	6647.6256
93	46.5	292.1688	6792.9246
94	47.0	295.3104	6939.7944
95	47.5	298.4520	7088.2350
96	48.0	301.5936	7238.2464
97	48.5	304.7352	7389.8286
98	49.0	307.8768	7542.9816
99	49.5	311.0184	7697.7054
100	50.0	314.1600	7854.0000

TRIGONOMETRIC FUNCTIONS

Degree	Sin	Cos	Tan	Cot	Sec	Csc
0	0.0000	1.0000	0.0000	Infinity	1.0000	Infinity
½	0.0087	1.0000	0.0087	114.5887	1.0000	114.5930
1	0.0175	0.9998	0.0175	57.2900	1.0002	57.2987
1½	0.0262	0.9997	0.0262	38.1885	1.0003	38.2016
2	0.0349	0.9994	0.0349	28.6363	1.0006	28.6537
2½	0.0436	0.9990	0.0437	22.9038	1.0010	22.9256
3	0.0523	0.9986	0.0524	19.0811	1.0014	19.1073
3½	0.0610	0.9981	0.0612	16.3499	1.0019	16.3804
4	0.0698	0.9976	0.0699	14.3007	1.0024	14.3356
4½	0.0785	0.9969	0.0787	12.7062	1.0031	12.7455
5	0.0872	0.9962	0.0875	11.4301	1.0038	11.4737
5½	0.0958	0.9954	0.0963	10.3854	1.0046	10.4334
6	0.1045	0.9945	0.1051	9.5144	1.0055	9.5668
6½	0.1132	0.9936	0.1139	8.7769	1.0065	8.8337
7	0.1219	0.9925	0.1228	8.1443	1.0075	8.2055
7½	0.1305	0.9914	0.1317	7.5958	1.0086	7.6613
8	0.1392	0.9903	0.1405	7.1154	1.0098	7.1853
8½	0.1478	0.9890	0.1495	6.6912	1.0111	6.7655
9	0.1564	0.9877	0.1584	6.3138	1.0125	6.3925
9½	0.1650	0.9863	0.1673	5.9758	1.0139	6.0589
10	0.1736	0.9848	0.1763	5.6713	1.0154	5.7588
10½	0.1822	0.9833	0.1853	5.3955	1.0170	5.4874
11	0.1908	0.9816	0.1944	5.1446	1.0187	5.2408

TRIGONOMETRIC FUNCTIONS *(cont.)*					
Degree	**Sin**	**Cos**	**Tan**	**Cot**	**Sec**	**Csc**
11½	0.1994	0.9799	0.2035	4.9152	1.0205	5.0159
12	0.2079	0.9781	0.2126	4.7046	1.0223	4.8097
12½	0.2164	0.9763	0.2217	4.5107	1.0243	4.6202
13	0.2250	0.9744	0.2309	4.3315	1.0263	4.4454
13½	0.2334	0.9724	0.2401	4.1653	1.0284	4.2837
14	0.2419	0.9703	0.2493	4.0108	1.0306	4.1336
14½	0.2504	0.9681	0.2586	3.8667	1.0329	3.9939
15	0.2588	0.9659	0.2679	3.7321	1.0353	3.8637
15½	0.2672	0.9636	0.2773	3.6059	1.0377	3.7420
16	0.2756	0.9613	0.2867	3.4874	1.0403	3.6280
16½	0.2840	0.9588	0.2962	3.3759	1.0429	3.5209
17	0.2924	0.9563	0.3057	3.2709	1.0457	3.4203
17½	0.3007	0.9537	0.3153	3.1716	1.0485	3.3255
18	0.3090	0.9511	0.3249	3.0777	1.0515	3.2361
18½	0.3173	0.9483	0.3346	2.9887	1.0545	3.1515
19	0.3256	0.9455	0.3443	2.9042	1.0576	3.0716
19½	0.3338	0.9426	0.3541	2.8239	1.0608	2.9957
20	0.3420	0.9397	0.3640	2.7475	1.0642	2.9238
20½	0.3502	0.9367	0.3739	2.6746	1.0676	2.8555
21	0.3584	0.9336	0.3839	2.6051	1.0711	2.7904
21½	0.3665	0.9304	0.3939	2.5386	1.0748	2.7285
22	0.3746	0.9272	0.4040	2.4751	1.0785	2.6695
22½	0.3827	0.9239	0.4142	2.4142	1.0824	2.6131

TRIGONOMETRIC FUNCTIONS (cont.)

Degree	Sin	Cos	Tan	Cot	Sec	Csc
23	0.3907	0.9205	0.4245	2.3559	1.0864	2.5593
23½	0.3987	0.9171	0.4348	2.2998	1.0904	2.5078
24	0.4067	0.9135	0.4452	2.2460	1.0946	2.4586
24½	0.4147	0.9100	0.4557	2.1943	1.0989	2.4114
25	0.4226	0.9063	0.4663	2.1445	1.1034	2.3662
25½	0.4305	0.9026	0.4770	2.0965	1.1079	2.3228
26	0.4384	0.8988	0.4877	2.0503	1.1126	2.2812
26½	0.4462	0.8949	0.4986	2.0057	1.1174	2.2412
27	0.4540	0.8910	0.5095	1.9626	1.1223	2.2027
27½	0.4617	0.8870	0.5206	1.9210	1.1274	2.1657
28	0.4695	0.8829	0.5317	1.8807	1.1326	2.1301
28½	0.4772	0.8788	0.5430	1.8418	1.1379	2.0957
29	0.4848	0.8746	0.5543	1.8040	1.1434	2.0627
29½	0.4924	0.8704	0.5658	1.7675	1.1490	2.0308
30	0.5000	0.8660	0.5774	1.7321	1.1547	2.0000
30½	0.5075	0.8616	0.5890	1.6977	1.1606	1.9703
31	0.5150	0.8572	0.6009	1.6643	1.1666	1.9416
31½	0.5225	0.8526	0.6128	1.6319	1.1728	1.9139
32	0.5299	0.8480	0.6249	1.6003	1.1792	1.8871
32½	0.5373	0.8434	0.6371	1.5697	1.1857	1.8612
33	0.5446	0.8387	0.6494	1.5399	1.1924	1.8361
33½	0.5519	0.8339	0.6619	1.5108	1.1992	1.8118
34	0.5592	0.8290	0.6745	1.4826	1.2062	1.7883

TRIGONOMETRIC FUNCTIONS *(cont.)*					
Degree	Sin	Cos	Tan	Cot	Sec	Csc
34½	0.5664	0.8241	0.6873	1.4550	1.2134	1.7655
35	0.5736	0.8192	0.7002	1.4281	1.2208	1.7434
35½	0.5807	0.8141	0.7133	1.4019	1.2283	1.7221
36	0.5878	0.8090	0.7265	1.3764	1.2361	1.7013
36½	0.5948	0.8039	0.7400	1.3514	1.2440	1.6812
37	0.6018	0.7986	0.7536	1.3270	1.2521	1.6616
37½	0.6088	0.7934	0.7673	1.3032	1.2605	1.6427
38	0.6157	0.7880	0.7813	1.2799	1.2690	1.6243
38½	0.6225	0.7826	0.7954	1.2572	1.2778	1.6064
39	0.6293	0.7771	0.8098	1.2349	1.2868	1.5890
39½	0.6361	0.7716	0.8243	1.2131	1.2960	1.5721
40	0.6428	0.7660	0.8391	1.1918	1.3054	1.5557
40½	0.6494	0.7604	0.8541	1.1708	1.3151	1.5398
41	0.6561	0.7547	0.8693	1.1504	1.3250	1.5243
41½	0.6626	0.7490	0.8847	1.1303	1.3352	1.5092
42	0.6691	0.7431	0.9004	1.1106	1.3456	1.4945
42½	0.6756	0.7373	0.9163	1.0913	1.3563	1.4802
43	0.6820	0.7314	0.9325	1.0724	1.3673	1.4663
43½	0.6884	0.7254	0.9490	1.0538	1.3786	1.4527
44	0.6947	0.7193	0.9657	1.0355	1.3902	1.4396
44½	0.7009	0.7133	0.9827	1.0176	1.4020	1.4267
45	0.7071	0.7071	1.0000	1.0000	1.4142	1.4142
45½	0.7133	0.7009	1.0176	0.9827	1.4267	1.4020

Degree	Sin	Cos	Tan	Cot	Sec	Csc
46	0.7193	0.6947	1.0355	0.9657	1.4396	1.3902
46½	0.7254	0.6884	1.0538	0.9490	1.4527	1.3786
47	0.7314	0.6820	1.0724	0.9325	1.4663	1.3673
47½	0.7373	0.6756	1.0913	0.9163	1.4802	1.3563
48	0.7431	0.6691	1.1106	0.9004	1.4945	1.3456
48½	0.7490	0.6626	1.1303	0.8847	1.5092	1.3352
49	0.7547	0.6561	1.1504	0.8693	1.5243	1.3250
49½	0.7604	0.6494	1.1708	0.8541	1.5398	1.3151
50	0.7660	0.6428	1.1918	0.8391	1.5557	1.3054
50½	0.7716	0.6361	1.2131	0.8243	1.5721	1.2960
51	0.7771	0.6293	1.2349	0.8098	1.5890	1.2868
51½	0.7826	0.6225	1.2572	0.7954	1.6064	1.2778
52	0.7880	0.6157	1.2799	0.7813	1.6243	1.2690
52½	0.7934	0.6088	1.3032	0.7673	1.6427	1.2605
53	0.7986	0.6018	1.3270	0.7536	1.6616	1.2521
53½	0.8039	0.5948	1.3514	0.7400	1.6812	1.2440
54	0.8090	0.5878	1.3764	0.7265	1.7013	1.2361
54½	0.8141	0.5807	1.4019	0.7133	1.7221	1.2283
55	0.8192	0.5736	1.4281	0.7002	1.7434	1.2208
55½	0.8241	0.5664	1.4550	0.6873	1.7655	1.2134
56	0.8290	0.5592	1.4826	0.6745	1.7883	1.2062
56½	0.8339	0.5519	1.5108	0.6619	1.8118	1.1992
57	0.8387	0.5446	1.5399	0.6494	1.8361	1.1924

TRIGONOMETRIC FUNCTIONS *(cont.)*					
Degree	**Sin**	**Cos**	**Tan**	**Cot**	**Sec**	**Csc**
57½	0.8434	0.5373	1.5697	0.6371	1.8612	1.1857
58	0.8480	0.5299	1.6003	0.6249	1.8871	1.1792
58½	0.8526	0.5225	1.6319	0.6128	1.9139	1.1728
59	0.8572	0.5150	1.6643	0.6009	1.9416	1.1666
59½	0.8616	0.5075	1.6977	0.5890	1.9703	1.1606
60	0.8660	0.5000	1.7321	0.5774	2.0000	1.1547
60½	0.8704	0.4924	1.7675	0.5658	2.0308	1.1490
61	0.8746	0.4848	1.8040	0.5543	2.0627	1.1434
61½	0.8788	0.4772	1.8418	0.5430	2.0957	1.1379
62	0.8829	0.4695	1.8807	0.5317	2.1301	1.1326
62½	0.8870	0.4617	1.9210	0.5206	2.1657	1.1274
63	0.8910	0.4540	1.9626	0.5095	2.2027	1.1223
63½	0.8949	0.4462	2.0057	0.4986	2.2412	1.1174
64	0.8988	0.4384	2.0503	0.4877	2.2812	1.1126
64½	0.9026	0.4305	2.0965	0.4770	2.3228	1.1079
65	0.9063	0.4226	2.1445	0.4663	2.3662	1.1034
65½	0.9100	0.4147	2.1943	0.4557	2.4114	1.0989
66	0.9135	0.4067	2.2460	0.4452	2.4586	1.0946
66½	0.9171	0.3987	2.2998	0.4348	2.5078	1.0904
67	0.9205	0.3907	2.3559	0.4245	2.5593	1.0864
67½	0.9239	0.3827	2.4142	0.4142	2.6131	1.0824
68	0.9272	0.3746	2.4751	0.4040	2.6695	1.0785
68½	0.9304	0.3665	2.5386	0.3939	2.7285	1.0748

TRIGONOMETRIC FUNCTIONS *(cont.)*						
Degree	Sin	Cos	Tan	Cot	Sec	Csc
69	0.9336	0.3584	2.6051	0.3839	2.7904	1.0711
69½	0.9367	0.3502	2.6746	0.37388	2.8555	1.0676
70	0.9397	0.3420	2.7475	0.36397	2.9238	1.0642
70½	0.9426	0.3338	2.8239	0.35412	2.9957	1.0608
71	0.9455	0.3256	2.9042	0.34433	3.0716	1.0576
71½	0.9483	0.3173	2.9887	0.33460	3.1515	1.0545
72	0.9511	0.3090	3.0777	0.32492	3.2361	1.0515
72½	0.9537	0.3007	3.1716	0.31530	3.3255	1.0485
73	0.9563	0.2924	3.2709	0.30573	3.4203	1.0457
73½	0.9588	0.2840	3.3759	0.29621	3.5209	1.0429
74	0.9613	0.2756	3.4874	0.28675	3.6280	1.0403
74½	0.9636	0.2672	3.6059	0.27732	3.7420	1.0377
75	0.9659	0.2588	3.7321	0.26795	3.8637	1.0353
75½	0.9681	0.2504	3.8667	0.25862	3.9939	1.0329
76	0.9703	0.2419	4.0108	0.24933	4.1336	1.0306
76½	0.9724	0.2334	4.1653	0.24008	4.2837	1.0284
77	0.9744	0.2250	4.3315	0.23087	4.4454	1.0263
77½	0.9763	0.2164	4.5107	0.22169	4.6202	1.0243
78	0.9781	0.2079	4.7046	0.21256	4.8097	1.0223
78½	0.9799	0.1994	4.9152	0.20345	5.0159	1.0205
79	0.9816	0.1908	5.1446	0.19438	5.2408	1.0187
79½	0.9833	0.1822	5.3955	0.18534	5.4874	1.0170

TRIGONOMETRIC FUNCTIONS *(cont.)*					
Degree	Sin	Cos	Tan	Cot	Sec	Csc
80	0.9848	0.1736	5.6713	0.17633	5.7588	1.0154
80½	0.9863	0.1650	5.9758	0.16734	6.0589	1.0139
81	0.9877	0.1564	6.3138	0.15838	6.3925	1.0125
81½	0.9890	0.1478	6.6912	0.14945	6.7655	1.0111
82	0.9903	0.1392	7.1154	0.14054	7.1853	1.0098
82½	0.9914	0.1305	7.5958	0.13165	7.6613	1.0086
83	0.9925	0.1219	8.1443	0.12278	8.2055	1.0075
83½	0.9936	0.1132	8.7769	0.11394	8.8337	1.0065
84	0.9945	0.1045	9.5144	0.10510	9.5668	1.0055
84½	0.9954	0.0958	10.3854	0.09629	10.4334	1.0046
85	0.9962	0.0872	11.4301	0.08749	11.4737	1.0038
85½	0.9969	0.0785	12.7062	0.07870	12.7455	1.0031
86	0.9976	0.0698	14.3007	0.06993	14.3356	1.0024
86½	0.9981	0.0610	16.3499	0.06116	16.3804	1.0019
87	0.9986	0.0523	19.0811	0.05241	19.1073	1.0014
87½	0.9990	0.0436	22.9038	0.04366	22.9256	1.0010
88	0.9994	0.0349	28.6363	0.03492	28.6537	1.0006
88½	0.9997	0.0262	38.1885	0.02619	38.2016	1.0003
89	0.9998	0.0175	57.2900	0.01746	57.2987	1.0002
89½	1.0000	0.0087	114.5887	0.00873	114.5930	1.0000
90	1.0000	0.0000	Infinity	0.00000	Infinity	1.0000

MEASURING BUILDING GROSS AREA

Measuring Procedure

1. Decide your preferred rule to measure areas: from exterior surface, center line, or inside surface of building perimeter wall.
2. Find out how many individual floors the job has (basement, main floor, upper floors, loft, penthouse, etc.).
3. Subdivide each floor into smaller segments that are easy to measure.
4. Measure each segment and summarize them to get the area for each floor.
5. Add up each floor to get the total building gross area.

Areas to Exclude or Keep Separate

- Pipe basement or crawl space
- Areaways either grilled over or open
- Cat-Walks
- Outside ramps or steps (without cover)
- Mezzanine or balcony when as a grille floor without equipment
- Porches
- Outside balconies
- Areaways
- Leading platforms
- Covered driveways
- Other roofed areas or passage without enclosing walls
- Cooling towers
- Exposed mechanical equipment enclosed with a screen wall but not roofed
- Fuel tanks or pneumatic tanks placed underground
- Oxygen storage tanks placed on a slab at ground level enclosed by a fence or screen

MASTERFORMAT 1995 EDITION

Series 0 — Bidding and contracting requirements

Division 01 General Requirements

Division 02 Site Work

Division 03 Concrete

Division 04 Masonry

Division 05 Metals

Division 06 Wood, Plastics

Division 07 Thermal and Moisture Protection

Division 08 Doors and Windows

Division 09 Finishes

Division 10 Specialties

Division 11 Equipment

Division 12 Furnishings

Division 13 Special Construction

Division 14 Conveying Equipment

Division 15 Mechanical

Division 16 Electrical

MASTERFORMAT 1995 EDITION HVAC TITLES

Main Titles

Division 15 Mechanical

15050 Basic Mechanical Materials and Methods

15100 Building Service Piping

15200 Process Piping

15300 Fire Protection Piping
      (By Fire Protection Contractor)

15400 Plumbing Fixtures and Equipment
      (By Plumber)

15500 Heat-Generation Equipment

15600 Refrigeration Equipment

15700 Heating, Ventilating, and Air Conditioning
      Equipment

15800 Air Distribution

15900 HVAC Instrumentation and Controls

15950 Testing, Adjusting, and Balancing

MASTERFORMAT 1995 EDITION HVAC TITLES *(cont.)*

Detailed Breakdown

Code	Short Description	Extended Description
15050	**Basic Mechanical Materials and Method**	
15060	Hangers and Support	
15070	Mechanical, Sound, Vibration, and Seismic Control	
15075	Mechanical Identification	
15080	Mechanical Insulation	Duct Insulation
		Equipment Insulation
		Piping Insulation
15090	Mechanical Restoration and Retrofit	
15100	**Building Services Piping**	
15105	Pipes and Tubes	
15110	Valve	
15120	Piping Specialties	
15130	Pump	

MASTERFORMAT 1995 EDITION HVAC TITLES *(cont.)*

Detailed Breakdown

Code	Short Description	Extended Description
15140	Domestic Water Piping	Disinfecting Potable Water Piping
		Non - Potable Water Piping
		Potable Water Piping
15150	Sanitary Waste and Vent Piping	Interceptors
		Sanitary Piping
		Sanitary Piping Separators
15160	Storm Drainage Piping	Retrofit Roof Drains
		Roof Drains
		Roof Drain Specialties
		Storm Drainage Piping
15170	Swimming Pool and Fountain Piping	Reflecting Pool and Fountain Piping
		Reflecting Pool and Fountain Specialties
		Swimming Pool Piping
		Swimming Pool Specialties

MASTERFORMAT 1995 EDITION HVAC TITLES *(cont.)*

Detailed Breakdown

Code	Short Description	Extended Description
15180	Heating and Cooling Piping	Condensate Drain Piping
		Heating and Cooling Pumps
		Hydronic Piping
		Refrigerant Piping
		Steam and Condensate Piping
		Steam Condensate Pumps
		Water Treatment Equipment
15190	Fuel Piping	Fuel Oil Piping
		Fuel Oil Pumps
		Fuel Oil Specialties
		Liquid Petroleum Gas Piping
		Natural Gas and Liquid Petroleum Gas Specialties
		Natural Gas Piping

MASTERFORMAT 1995 EDITION HVAC TITLES (cont.)

Detailed Breakdown

Code	Short Description	Extended Description
15200	**Process Piping**	
15210	Process Air and Gas Piping	Air Compressors
		Compressed Air Piping
		Gas Equipment
		Gas Piping
		Helium Piping
		Nitrogen Piping
		Nitrous Oxide Gas Piping
		Vacuum Pumps
		Vacuum Piping
15220	Process Water and Waste Piping	Deionized Water Piping
		Distilled Water Piping
		Laboratory Acid Waste and Vent Piping
		Process Piping Interceptors
		Reverse Osmosis Water Piping

MASTERFORMAT 1995 EDITION HVAC TITLES (cont.)

Detailed Breakdown

Code	Short Description	Extended Description
15230	Industrial Process Piping	Dry Product Piping
		Fluid Product Piping
15300	Fire Protection Piping	
15400	Plumbing Fixtures and Equipment	
15410	Plumbing Fixture	
15440	Plumbing Pump	Base-Mounted Pumps
		Compact Circulators
		Inline Pumps
		Packaged Booster Pumping Station
		Sewage Ejectors
		Sump Pump
15450	Potable Water Storage Tank	
15460	Domestic Water Conditioning Equipment	Water Conditioners
		Water Softeners
15470	Domestic Water Filtrating Equipment	Disposable Filters
		Rechargeable Filters

MASTERFORMAT 1995 EDITION HVAC TITLES *(cont.)*

Detailed Breakdown

Code	Short Description	Extended Description
15480	Domestic Water Heater	Domestic Water Heat Exchangers
		Packaged Domestic Water Heaters
15490	Pool and Fountain Equipment	Fountain Equipment
		Reflecting Pool Equipment
		Swimming Pool Equipment
15500	**Heat-Generation Equipment**	
15510	Heating Boilers and Accessories	Cast-Iron Boilers
		Condensing Boilers
		Finned Water-Tube Boilers
		Firebox Heating Boilers
		Flexible Water-Tube Boilers
		Pulse Combustion Boilers
		Scotch Marine Boilers
		Steel Water-Tube Boilers

MASTERFORMAT 1995 EDITION HVAC TITLES *(cont.)*

Detailed Breakdown

Code	Short Description	Extended Description
15520	Feedwater Equipment	Boiler Feed Water Pumps
		Deaerators
		Packaged Deaerator and Feedwater Pump
15530	Furnace	Electric-Resistance Furnaces
		Fuel-Fired Furnaces
15540	Fuel-Fired Heater	Fuel-Fired Duct Heaters
		Fuel-Fired Radiant Heaters
		Fuel-Fired Unit Heaters
15550	Breechings, Chimneys, and Stack	Draft Control Devices
		Fabricated Breechings and Accessories
		Fabricated Stacks
		Gas Vents
		Insulated Sectional Chimneys

MASTERFORMAT 1995 EDITION HVAC TITLES (cont.)

Detailed Breakdown

Code	Short Description	Extended Description
15600	**Refrigeration Equipment**	
15610	Refrigeration Compressor	Centrifugal Refrigerant Compressors
		Reciprocating Refrigerant Compressors
		Rotary-Screw Refrigerant Compressors
15620	Packaged Water Chiller	Absorption Water Chillers
		Centrifugal Water Chillers
		Reciprocating Water Chillers
		Rotary-Screw Water Chillers
15630	Refrigerant Monitoring System	
15640	Packaged Cooling Tower	Mechanical-Draft Cooling Towers
		Natural-Draft Cooling Towers
15650	Field-Erected Cooling Tower	
15660	Liquid Coolers and Evaporative Condenser	
15670	Refrigerant Condensing Unit	Packaged Refrigerant Condensing Coils and Fan Units
		Refrigerant Condensing Coils

MASTERFORMAT 1995 EDITION HVAC TITLES *(cont.)*

Detailed Breakdown

Code	Short Description	Extended Description
15700	Heating, Ventilating, and Air Conditioning Equipment	
15710	Heat Exchanger	Steam-to-Water Heat Exchanges
		Water-to-Water Heat Exchanges
15720	Air Handling Unit	Built-Up Indoor Air Handling Units
		Customized Rooftop Air Handling Units
		Modular Indoor Air Handling Units
		Modular Rooftop Air Handling Units
15730	Unitary Air Conditioning Equipment	Packaged Air Conditioners
		Packaged Rooftop Air Conditioning Units
		Packaged Rooftop Make-Up Air Conditioning Units
		Packaged Terminal Air Conditioning Units
		Room Air Conditioners
		Split System Air Conditioning Units

MASTERFORMAT 1995 EDITION HVAC TITLES (cont.)

Detailed Breakdown

Code	Short Description	Extended Description
15740	Heat Pump	Air-Source Heat Pumps
		Rooftop Heat Pumps
		Water-Source Heat Pumps
15750	Humidity Control Equipment	Dehumidifiers
		Humidifiers
		Swimming Pool Dehumidification Units
15760	Terminal Heating and Cooling Unit	Air Coils
		Convectors
		Fan Coil Units
		Finned-Tube Radiation
		Induction Units
		Infrared Heaters
		Unit Heaters
		Unit Ventilators
15770	Floor-Heating and Snow-Melting Equipment	

MASTERFORMAT 1995 EDITION HVAC TITLES *(cont.)*

Detailed Breakdown

Code	Short Description	Extended Description
15780	Energy Recovery Equipment	Energy Storage Tanks
		Heat Pipe
		Heat Wheels
15800	**Air Distribution**	
15810	Duct	Duct Hangers and Supports
		Fibrous Glass Ducts
		Flexible Ducts
		Glass-Fiber-Reinforced Plastic Ducts
		Metal Ducts
15820	Duct Accessories	
15830	Fan	Air Curtains Axial Fans
		Ceiling Fans
		Centrifugal Fans
		Industrial Ventilating Equipment
		Power Ventilators

MASTERFORMAT 1995 EDITION HVAC TITLES (cont.)

Detailed Breakdown

Code	Short Description	Extended Description
15840	Air Terminal Unit	Constant Volume
		Variable Volume Units
15850	Air Outlets and Inlet	Diffusers, Registers, and Grilles
		Gravity Ventilators
		Intake and Relief Ventilators
		Penthouse Ventilators
15860	Air Cleaning Device	Air Filters
		Dust Collectors
		Electronic Air Cleaners
		High-Efficiency Air Filters
		ULPA Filters
15900	**HVAC Instrumentation and Control**	
15905	HVAC Instrumentation	
15910	Direct Digital Control	

MASTERFORMAT 1995 EDITION HVAC TITLES *(cont.)*

Detailed Breakdown

Code	Short Description	Extended Description
15915	Electric and Electronic Control	Central Control Equipment
		Control Panels
		Terminal Boxes
15920	Pneumatic Control	
15925	Pneumatic and Electric Control	
15930	Self-Powered Control	
15935	Building Systems Control	
15940	Sequence of Operation	
15950	Testing, Adjusting, Balancing	Demonstration of Mechanical Equipment
		Duct Testing, Adjusting, and Balancing
		Equipment Testing, Adjusting, and Balancing
		Mechanical Equipment Starting
		Pipe Testing, Adjusting, and Balancing

MASTERFORMAT 2004 EDITION

Note: In the new MasterFormat 2004 Edition, Division 23 covers most of the HVAC items originally under Division 15 in the old MasterFormat 1995 edition.

PROCUREMENT AND CONTRACTING REQUIREMENTS GROUP

Division 00 Procurement and Contracting Requirements

SPECIFICATIONS GROUP

General Requirements Subgroup

Division 01 General Requirements

Facility Construction Subgroup

Division 02 Existing Conditions

Division 03 Concrete

Division 04 Masonry

Division 05 Metals

Division 06 Wood, Plastics, and Composites

Division 07 Thermal and Moisture Protection

Division 08 Openings

Division 09 Finishes

Division 10 Specialties

Division 11 Equipment

Division 12 Furnishings

Division 13 Special Construction

MASTERFORMAT 2004 EDITION *(cont.)*

Division 14 Conveying Equipment

Division 15 Reserved

Division 16 Reserved

Division 17 Reserved

Division 18 Reserved

Division 19 Reserved

Facility Services Subgroup

Division 20 Reserved

Division 21 Fire Suppression

Division 22 Plumbing

Division 23 Heating, Ventilating, and Air
Conditioning

Division 24 Reserved

Division 25 Integrated Automation

Division 26 Electrical

Division 27 Communications

Division 28 Electronic Safety and Security

Division 29 Reserved

Site And Infrastructure Subgroup

Division 30 Reserved

Division 31 Earthwork

Division 32 Exterior Improvements

Division 33 Utilities

MASTERFORMAT 2004 EDITION *(cont.)*

Site And Infrastructure Subgroup *(cont.)*

Division 34 Transportation

Division 35 Waterway and Marine Construction

Division 36 Reserved

Division 37 Reserved

Division 38 Reserved

Division 39 Reserved

Process Equipment Subgroup

Division 40 Process Integration

Division 41 Material Processing and Handling
          Equipment

Division 42 Process Heating, Cooling, and Drying
          Equipment

Division 43 Process Gas and Liquid Handling,
          Purification, and Storage Equipment

Division 44 Pollution Control Equipment

Division 45 Industry-Specific Manufacturing
          Equipment

Division 46 Reserved

Division 47 Reserved

Division 48 Electrical Power Generation

Division 49 Reserved

MASTERFORMAT 2004 EDITION HVAC TITLES

Main Titles

23 00 00 Heating, Ventilating, and Air-conditioning (HVAC)

23 10 00 Facility Fuel Systems

23 20 00 HVAC Piping and Pumps

23 30 00 HVAC Air Distribution

23 40 00 HVAC Air Cleaning Devices

23 50 00 Central Heating Equipment

23 60 00 Central Cooling Equipment

23 70 00 Central HVAC Equipment

23 80 00 Decentralized HVAC Equipment

23 90 00 Unassigned

Detailed Breakdown

23 00 00 HEATING, VENTILATING, AND AIR-CONDITIONING (HVAC)

23 01 00 Operation and Maintenance of HVAC Systems

23 01 10 Operation and Maintenance of Facility Fuel Systems

23 01 20 Operation and Maintenance of HVAC Piping and Pumps

23 01 30 Operation and Maintenance of HVAC Air Distribution

  23 01 30.51 HVAC Air Duct Cleaning

23 01 50 Operation and Maintenance of Central Heating Equipment

23 01 60 Operation and Maintenance of Central Cooling Equipment

23 01 60.71 Refrigerant Recovery/Recycling

23 01 70 Operation and Maintenance of Central HVAC Equipment

23 01 80 Operation and Maintenance of Decentralized HVAC Equipment

23 01 90 Diagnostic Systems for HVAC

23 05 00 Common Work Results for HVAC

23 05 13 Common Motor Requirements for HVAC Equipment

23 05 16 Expansion Fittings and Loops for HVAC Piping

23 05 19 Meters and Gages for HVAC Piping

23 05 23 General-Duty Valves for HVAC Piping

23 05 29 Hangers and Supports for HVAC Piping and Equipment

23 05 33 Heat Tracing for HVAC Piping

23 05 48 Vibration and Seismic Controls for HVAC Piping and Equipment

23 05 53 Identification for HVAC Piping and Equipment

23 05 63 Anti-Microbial Coatings for HVAC Ducts and Equipment

23 05 66 Anti-Microbial Ultraviolet Emitters for HVAC Ducts and Equipment

23 05 93 Testing, Adjusting, and Balancing for HVAC

23 06 00 Schedules for HVAC

23 06 10 Schedules for Facility Fuel Service Systems

23 06 20 Schedules for HVAC Piping and Pumps

23 06 20.13 Hydronic Pump Schedule

23 06 30 Schedules for HVAC Air Distribution

23 06 30.13 HVAC Fan Schedule

23 06 30.16 Air Terminal Unit Schedule

23 06 30.19 Air Outlet and Inlet Schedule

23 06 30.23 HVAC Air Cleaning Device Schedule

23 06 50 Schedules for Central Heating Equipment

23 06 50.13 Heating Boiler Schedule

23 06 60 Schedules for Central Cooling Equipment

23 06 60.13 Refrigerant Condenser Schedule

23 06 60.16 Packaged Water Chiller Schedule

23 06 70 Schedules for Central HVAC Equipment

23 06 70.13 Indoor, Central-Station Air-Handling Unit Schedule

23 06 70.16 Packaged Outdoor HVAC Equipment Schedule

23 06 80 Schedules for Decentralized HVAC Equipment

23 06 80.13 Decentralized Unitary HVAC Equipment Schedule

MASTERFORMAT 2004 EDITION HVAC TITLES *(cont.)*

23 13 00 Facility Fuel-Storage Tanks

23 13 13 Facility Underground Fuel-Oil, Storage Tanks

23 13 13.13 Double- Wall Steel, Underground Fuel-Oil, Storage Tanks

23 13 13.16 Composite, Steel, Underground Fuel-Oil, Storage Tanks

23 13 13.19 Jacketed, Steel, Underground Fuel-Oil, Storage Tanks

23 13 13.23 Glass-Fiber-Reinforced-Plastic, Underground Fuel-Oil, Storage Tanks

23 13 13.33 Fuel-Oil Storage Tank Pumps

23 13 23 Facility Aboveground Fuel-Oil, Storage Tanks

23 13 23.13 Vertical, Steel, Aboveground Fuel-Oil, Storage Tanks

23 13 23.16 Horizontal, Steel, Aboveground Fuel-Oil, Storage Tanks

23 13 23.19 Containment-Dike, Steel, Aboveground Fuel-Oil, Storage Tanks

23 13 23.23 Insulated, Steel, Aboveground Fuel-Oil, Storage Tanks

23 13 23.26 Concrete-Vaulted, Steel, Aboveground Fuel-Oil, Storage Tanks

23 20 00 HVAC PIPING AND PUMPS

23 21 00 Hydronic Piping and Pumps

23 21 13 Hydronic Piping

23 21 13.13 Underground Hydronic Piping

23 21 13.23 Aboveground Hydronic Piping

23 21 13.33 Ground-Loop Heat-Pump Piping

23 21 23 Hydronic Pumps

23 21 23.13 In-Line Centrifugal Hydronic Pumps

MASTERFORMAT 2004 EDITION HVAC TITLES *(cont.)*

23 21 23.16 Base-Mounted, Centrifugal Hydronic Pumps

23 21 23.19 Vertical-Mounted, Double-Suction Centrifugal Hydronic Pumps

23 21 23.23 Vertical-Turbine Hydronic Pumps

23 21 29 Automatic Condensate Pump Units

23 22 00 Steam and Condensate Piping and Pumps

23 22 13 Steam and Condensate Heating Piping

23 22 13.13 Underground Steam and Condensate Heating Piping

23 22 13.23 Aboveground Steam and Condensate Heating Piping

23 22 23 Steam Condensate Pumps

23 22 23.13 Electric-Driven Steam Condensate Pumps

23 22 23.23 Pressure-Powered Steam Condensate Pumps

23 23 00 Refrigerant Piping

23 23 13 Refrigerant Piping Valves

23 23 16 Refrigerant Piping Specialties

23 23 19 Refrigerant Safety Relief Valve Discharge Piping

23 23 23 Refrigerants

23 24 00 Internal-Combustion Engine Piping

23 24 13 Internal-Combustion Engine Remote-Radiator Coolant Piping

23 24 16 Internal-Combustion Engine Exhaust Piping

23 25 00 HVAC Water Treatment

23 25 13 Water Treatment for Closed-Loop Hydronic Systems

23 25 16 Water Treatment for Open Hydronic Systems

23 25 19 Water Treatment for Steam System Feedwater

23 25 23 Water Treatment for Humidification Steam System Feedwater

23 30 00 HVAC AIR DISTRIBUTION

23 31 00 HVAC Ducts and Casings

23 31 13 Metal Ducts

    23 31 13.13 Rectangular Metal Ducts

    23 31 13.16 Round and Flat-Oval Spiral Ducts

    23 31 13.19 Metal Duct Fittings

23 31 16 Nonmetal Ducts

    23 31 16.13 Fibrous-Glass Ducts

    23 31 16.16 Thermoset Fiberglass-Reinforced Plastic Ducts

    23 31 16.19 PVC Ducts

    23 31 16.26 Concrete Ducts

23 31 19 HVAC Casings

23 32 00 Air Plenums and Chases

23 32 13 Fabricated, Metal Air Plenums

23 32 33 Air-Distribution Ceiling Plenums

23 32 36 Air-Distribution Floor Plenums

23 32 39 Air-Distribution Wall Plenums

23 32 43 Air-Distribution Chases Formed by General Construction

23 32 48 Acoustical Air Plenums

23 33 00 Air Duct Accessories

23 33 13 Dampers

    23 33 13.13 Volume-Control Dampers

    23 33 13.16 Fire Dampers

    23 33 13.19 Smoke-Control Dampers

    23 33 13.23 Backdraft Dampers

23 33 19 Duct Silencers

23 33 23 Turning Vanes

23 33 33 Duct-Mounting Access Doors

23 33 43 Flexible Connectors

23 33 46 Flexible Ducts

23 33 53 Duct Liners

23 34 00 HVAC Fans

23 34 13 Axial HVAC Fans

23 34 16 Centrifugal HVAC Fans

23 34 23 HVAC Power Ventilators

23 34 33 Air Curtains

23 35 00 Special Exhaust Systems

23 35 13 Sawdust Collection Systems

23 35 16 Engine Exhaust Systems

23 35 16.13 Positive-Pressure Engine Exhaust Systems

23 35 16.16 Mechanical Engine Exhaust Systems

23 36 00 Air Terminal Units

23 36 13 Constant-Air-Volume Units

23 36 16 Variable-Air-Volume Units

23 37 00 Air Outlets and Inlets

23 37 13 Diffusers, Registers, and Grilles

23 37 16 Fabric Air Distribution Devices

23 37 23 HVAC Gravity Ventilators

23 37 23.13 HVAC Gravity Dome Ventilators

23 37 23.16 HVAC Gravity Louvered-Penthouse Ventilators

23 37 23.19 HVAC Gravity Upblast Ventilators

23 38 00 Ventilation Hoods

23 38 13 Commercial-Kitchen Hoods

23 38 13.13 Listed Commercial-Kitchen Hoods

23 38 13.16 Standard Commercial-Kitchen Hoods

23 38 16 Fume Hoods

23 40 00 HVAC AIR CLEANING DEVICES

23 41 00 Particulate Air Filtration

23 41 13 Panel Air Filters

23 41 16 Renewable-Media Air Filters

23 41 19 Washable Air Filters

23 41 23 Extended Surface Filters

23 41 33 High-Efficiency Particulate Filtration

23 41 43 Ultra-Low Penetration Filtration

23 41 46 Super Ultra-Low Penetration Filtration

23 42 00 Gas-Phase Air Filtration

23 42 13 Activated-Carbon Air Filtration

23 42 16 Chemically-Impregnated Adsorption Air Filtration

23 42 19 Catalytic-Adsorption Air Filtration

23 43 00 Electronic Air Cleaners

23 43 13 Washable Electronic Air Cleaners

23 43 16 Agglomerator Electronic Air Cleaners

23 43 23 Self-Contained Electronic Air Cleaners

23 50 00 CENTRAL HEATING EQUIPMENT

23 51 00 Breechings, Chimneys, and Stacks

23 51 13 Draft Control Devices

23 51 13.13 Draft-Induction Fans

23 51 13.16 Vent Dampers

23 51 13.19 Barometric Dampers

23 51 16 Fabricated Breechings and Accessories

23 51 19 Fabricated Stacks

23 51 23 Gas Vents

23 51 33 Insulated Sectional Chimneys

23 51 43 Flue-Gas Filtration Equipment

23 51 43.13 Gaseous Filtration

23 51 43.16 Particulate Filtration

23 52 00 Heating Boilers

23 52 13 Electric Boilers

23 52 16 Condensing Boilers

23 52 16.13 Stainless-Steel Condensing Boilers

23 52 16.16 Aluminum Condensing Boilers

23 52 19 Pulse Combustion Boilers

23 52 23 Cast-Iron Boilers

23 52 33 Water-Tube Boilers

23 52 33.13 Finned Water-Tube Boilers

23 52 33.16 Steel Water-Tube Boilers

23 52 33.19 Copper Water-Tube Boilers

23 52 39 Fire-Tube Boilers

23 52 39.13 Scotch Marine Boilers

23 52 39.16 Steel Fire-Tube Boilers

23 53 00 Heating Boiler Feedwater Equipment

23 53 13 Boiler Feedwater Pumps

23 53 16 Deaerators

23 54 00 Furnaces

23 54 13 Electric-Resistance Furnaces

23 54 16 Fuel-Fired Furnaces

23 54 16.13 Gas-Fired Furnaces

23 54 16.16 Oil-Fired Furnaces

23 55 00 Fuel-Fired Heaters

23 55 13 Fuel-Fired Duct Heaters

23 55 13.13 Oil-Fired Duct Heaters

23 55 13.16 Gas-Fired Duct Heaters

23 55 23 Gas-Fired Radiant Heaters

23 55 33 Fuel-Fired Unit Heaters

    23 55 33.13 Oil-Fired Unit Heaters

    23 55 33.16 Gas-Fired Unit Heaters

23 56 00 Solar Energy Heating Equipment

23 56 13 Heating Solar Collectors

    23 56 13.13 Heating Solar Flat-Plate Collectors

    23 56 13.16 Heating Solar Concentrating Collectors

    23 56 13.19 Heating Solar Vacuum-Tube Collectors

23 56 16 Packaged Solar Heating Equipment

23 57 00 Heat Exchangers for HVAC

23 57 13 Steam-to-Steam Heat Exchangers

23 57 16 Steam-to-Water Heat Exchangers

23 57 19 Liquid-to-Liquid Heat Exchangers

    23 57 19.13 Plate-Type, Liquid- to-Liquid Heat Exchangers

    23 57 19.16 Shell-Type, Liquid-to-Liquid Heat Exchangers

23 57 33 Direct Geoexchange Heat Exchangers

23 60 00 CENTRAL COOLING EQUIPMENT

23 61 00 Refrigerant Compressors

23 61 13 Centrifugal Refrigerant Compressors

    23 61 13.13 Non-Condensable Gas Purge Equipment

23 61 16 Reciprocating Refrigerant Compressors

23 61 19 Scroll Refrigerant Compressors

23 61 23 Rotary-Screw Refrigerant Compressors

23 62 00 Packaged Compressor and Condenser Units

23 62 13 Packaged Air-Cooled Refrigerant Compressor and Condenser Units

23 62 23 Packaged Water-Cooled Refrigerant Compressor and Condenser Units

23 63 00 Refrigerant Condensers

23 63 13 Air-Cooled Refrigerant Condensers

23 63 23 Water-Cooled Refrigerant Condensers

23 63 33 Evaporative Refrigerant Condensers

23 64 00 Packaged Water Chillers

23 64 13 Absorption Water Chillers

23 64 13.13 Direct-Fired Absorption Water Chillers

23 64 13.16 Indirect-Fired Absorption Water Chillers

23 64 16 Centrifugal Water Chillers

23 64 19 Reciprocating Water Chillers

23 64 23 Scroll Water Chillers

23 64 26 Rotary-Screw Water Chillers

23 65 00 Cooling Towers

23 65 13 Forced-Draft Cooling Towers

23 65 13.13 Open-Circuit, Forced-Draft Cooling Towers

23 65 13.16 Closed-Circuit, Forced-Draft Cooling Towers

23 65 16 Natural-Draft Cooling Towers

23 65 23 Field-Erected Cooling Towers

23 65 33 Liquid Coolers

23 70 00 CENTRAL HVAC EQUIPMENT

23 71 00 Thermal Storage

23 71 13 Thermal Heat Storage

23 71 13.13 Room Storage Heaters for Thermal Storage

23 71 13.16 Heat-Pump Boosters for Thermal Storage

23 71 13.19 Central Furnace Heat-Storage Units

23 71 13.23 Pressurized-Water Thermal Storage Tanks

23 71 16 Chilled-Water Thermal Storage

23 71 19 Ice Storage

    23 71 19.13 Internal Ice-on-Coil Thermal Storage

    23 71 19.16 External Ice-on-Coil Thermal Storage

    23 71 19.19 Encapsulated-Ice Thermal Storage

    23 71 19.23 Ice-Harvesting Thermal Storage

    23 71 19.26 Ice-Slurry Thermal Storage

23 72 00 Air-to-Air Energy Recovery Equipment

23 72 13 Heat-Wheel Air-to-Air Energy-Recovery Equipment

23 72 16 Heat-Pipe Air-to-Air Energy-Recovery Equipment

23 72 19 Fixed-Plate Air-to-Air Energy-Recovery Equipment

23 72 23 Packaged Air-to-Air Energy-Recovery Units

23 73 00 Indoor Central-Station Air-Handling Units

23 73 13 Modular Indoor Central-Station Air-Handling Units

23 73 23 Custom Indoor Central-Station Air-Handling Units

23 73 33 Indoor Indirect Fuel-Fired Heating and Ventilating Units

    23 73 33.13 Indoor Indirect Oil-Fired Heating and Ventilating Units

    23 73 33.16 Indoor Indirect Gas-Fired Heating and Ventilating Units

23 73 39 Indoor, Direct Gas-Fired Heating and Ventilating Units

23 74 00 Packaged Outdoor HVAC Equipment

23 74 13 Packaged, Outdoor, Central-Station Air-Handling Units

23 74 23 Packaged, Outdoor, Heating-Only Makeup-Air Units

MASTERFORMAT 2004 EDITION HVAC TITLES *(cont.)*

23 74 23.13 Packaged, Direct-Fired, Outdoor, Heating-Only Makeup-Air Units

23 74 23.16 Packaged, Indirect-Fired, Outdoor, Heating-Only Makeup-Air Units

23 74 33 Packaged, Outdoor, Heating and Cooling Makeup Air-Conditioners

23 75 00 Custom-Packaged Outdoor HVAC Equipment

23 75 13 Custom-Packaged, Outdoor, Central-Station Air-Handling Units

23 75 23 Custom-Packaged, Outdoor, Heating and Ventilating Makeup-Air Units

23 75 33 Custom-Packaged, Outdoor, Heating and Cooling Makeup Air-Conditioners

23 76 00 Evaporative Air-Cooling Equipment

23 76 13 Direct Evaporative Air Coolers

23 76 16 Indirect Evaporative Air Coolers

23 76 19 Combined Direct and Indirect Evaporative Air Coolers

23 80 00 DECENTRALIZED HVAC EQUIPMENT

23 81 00 Decentralized Unitary HVAC Equipment

23 81 13 Packaged Terminal Air-Conditioners

23 81 16 Room Air-Conditioners

23 81 19 Self-Contained Air-Conditioners

23 81 19.13 Small-Capacity Self-Contained Air-Conditioners

23 81 19.16 Large-Capacity Self-Contained Air-Conditioners

23 81 23 Computer-Room Air-Conditioners

23 81 26 Split-System Air-Conditioners

23 81 43 Air-Source Unitary Heat Pumps

23 81 46 Water-Source Unitary Heat Pumps

MASTERFORMAT 2004 EDITION HVAC TITLES *(cont.)*

23 82 00 Convection Heating and Cooling Units

23 82 13 Valance Heating and Cooling Units

23 82 16 Air Coils

23 82 19 Fan Coil Units

23 82 23 Unit Ventilators

23 82 26 Induction Units

23 82 29 Radiators

23 82 33 Convectors

23 82 36 Finned-Tube Radiation Heaters

23 82 39 Unit Heaters

    23 82 39.13 Cabinet Unit Heaters

    23 82 39.16 Propeller Unit Heaters

    23 82 39.19 Wall and Ceiling Unit Heaters

23 83 00 Radiant Heating Units

23 83 13 Radiant-Heating Electric Cables

    23 83 13.16 Radiant-Heating Electric Mats

23 83 16 Radiant-Heating Hydronic Piping

23 83 23 Radiant-Heating Electric Panels

23 83 33 Electric Radiant Heaters

23 84 00 Humidity Control Equipment

23 84 13 Humidifiers

23 84 16 Dehumidifiers

23 84 19 Indoor Pool and Ice-Rink Dehumidification Units

23 90 00 Unassigned

UNIFORMAT LEVELS AND TITLES

A Substructure
- A10   Foundations
- A20   Basement Construction

B Shell
- B10   Superstructure
- B20   Exterior Enclosure
- B30   Roofing

C Interiors
- C10   Interior Construction
- C20   Stairs
- C30   Interior Finishes

D Services
- D10   Conveying Systems
- D20   Plumbing
- D30   Heating, Ventilating, and Air Conditioning (HVAC)
- D40   Fire Protection Systems
- D50   Electrical Systems

E Equipment and Furnishings
- E10   Equipment
- E20   Furnishings

F Special Construction and Demolition
- F10   Special Construction
- F20   Selective Demolition

G Building Sitework
- G10   Site Preparation
- G20   Site Improvements
- G30   Site Civil/Mechanical Utilities
- G40   Site Electrical Utilities
- G90   Other Site Construction

Z General
- Z10   General Requirements
- Z20   Bidding Requirements, Contract Forms, and Conditions Contingencies
- Z90   Project Cost Estimate

Project Description
- 10   Project description
- 20   Proposal, bidding, and contracting
- 30   Cost summary

UNIFORMAT HVAC TITLES

D 30 HVAC

D 3010 Energy Supply

Includes	Excludes
• oil, gas, and coal supply	• electrical energy supply systems (see section D 5090 and D 5010)
• steam, hot, and chilled water supply	
• solar energy supply	
• wind energy supply	

D 3020 Heat Generating Systems

Includes	Excludes
• boilers, including electric	• electric space unit heaters and baseboard, fuel fired unit heaters, furnaces (see section D 3050)
• piping and fittings adjacent to boilers	• controls and instrumentation (see section D 3060)
• primary pumps	
• auxiliary equipment	
• equipment and piping insulation	

UNIFORMAT HVAC TITLES (cont.)

D 3030 Cooling Generating Systems

Includes	Excludes
• chillers	• secondary chilled water pumps (see section D 3040)
• cooling towers and evaporative coolers	• distribution piping (see section D 3040)
• condensing units	• controls/instrumentation (see section D 3060)
• piping and fittings	
• primary pumps	
• direct expansion systems	
• equipment and piping insulation	

D 3040 Distribution Systems

Includes	Excludes
• supply and return air systems, including air handling units with coils (electric included), filters, ductwork, and associated devices such as VAV boxes, duct heaters, induction units, and grilles	• electric, gas, or oil fired unit heaters (see section D 3050)
• ventilation and exhaust systems	• furnaces (gas or oil) (see section D 3050)
• steam, hot water, glycol, and chilled water distribution	• floor, ceiling, and rooftop package units (see section D 3050)

7-56

UNIFORMAT HVAC TITLES *(cont.)*

• associated terminal devices including convectors, fan-coil units, and induction, but not water and steam unit heaters	• controls and instrumentation (see section D 3060)
• heat recovery equipment	
• auxiliary equipment such as secondary pumps, heat exchangers, sound attenuation, and vibration isolation	
• piping, duct, and equipment insulation	

D 3050 Terminal and Package Units

Includes	Excludes
• electric baseboard	• piping and accessories (see section D 3040)
• electric or fossil fuel fired unit heaters, unit ventilators, and radiant heaters	• hydronic or steam convectors, fan-coil units (see section D 3040)
• window or through-the-wall air conditioners, with or without heating of any type	• cooling towers, remote air-cooled condensers, evaporative coolers (see section D 3030)
• reverse-cycle, water- or air-cooled, terminal heat pumps	• air-handling units with only hydronic heating or steam coils (see section D 3040)
• wall sleeves where required	• air-handling units with chilled water or direct expansion cooling coils (see section D 3040)

UNIFORMAT HVAC TITLES (cont.)

- electric or fossil fuel fired air-handling units or furnaces
- self-contained, air- or water-cooled, floor, ceiling, and rooftop air conditioners, and heat pumps
- ductwork and accessories, including flue stacks
- factory-integrated controls

D 3060 Controls and Instrumentation

Includes	Excludes
• heating generating systems	• factory-installed controls, when an integral part of terminal and package units (see section D 3050)
• cooling generating systems	
• heating/cooling air handling units	
• exhaust and ventilating systems	
• terminal devices	
• energy monitoring and control	
• building automation systems	

UNIFORMAT HVAC TITLES (cont.)

D 3070 Systems Testing and Balancing

Includes	Excludes
• piping systems testing and balancing	
• air systems testing and balancing	

D3090 Other HVAC Systems and Equipment

Includes	Excludes
• special cooling systems and devices	
• special humidity control	
• dust and fume collectors	
• air curtains	
• air purifiers	
• paint spray booth ventilation systems	
• general construction items associated with mechanical systems	

COMMON UNIT CONVERSIONS

Convert From	Multiply By	To Obtain
Acres	0.4047	Hectares
Acres	43,560	Square Feet
Acres	0.0016	Square Miles
Acres	4,047	Square Meters
Acres	0.0041	Square Kilometers
Acres	4840	Square Yards
Acre-feet	43560	Cubic Feet
Acre-feet	1,233	Cubic Meters
Acre-feet	1,613	Cubic Yards
Acre-feet	325,900	Gallons (US)
Acre-inches	3,630	Cubic Feet
Acre-inches	102.79	Cubic Meters
Acre-inches	134.44	Cubic Yards
Acre-inches	27,154	Gallons (US)
Atmospheres	76	Centimeters of Mercury
Atmospheres	29.92	Inches of Mercury
Atmospheres	1,033	Centimeters of Water
Atmospheres	33.9	Feet of Water
Atmospheres	101.325	Kilopascals
Atmospheres	101,325	Pascals
Atmospheres	14.7	Pounds per Square Inch
British Thermal Units	252.16	Calories

COMMON UNIT CONVERSIONS (cont.)

Convert From	Multiply By	To Obtain
British Thermal Units	778.17	Foot-Pounds
British Thermal Units	0.0004	Horsepower-hours
British Thermal Units	1,055	Joules
British Thermal Units	0.252	Kilogram-calories
British Thermal Units	107.51	Kilogram-meters
British Thermal Units	0.0003	Kilowatt-Hours
Calories	0.00397	British Thermal Units
Calories	4.184	Joules
Calories	3.086	Foot-Pounds
Centimeters	10	Millimeters
Centimeters	0.01	Meters
Centimeters	0.394	Inches
Centimeters	0.033	Feet
Centimeters	0.011	Yards
Cubic Feet	28.32	Liters
Cubic Feet	0.0283	Cubic Meters
Cubic Feet	1,728	Cubic Inches
Cubic Feet	957.51	Fluid Ounces (US)
Cubic Feet	59.844	Pints (US)
Cubic Feet	29.922	Quarts (US)
Cubic Feet	7.481	Gallons (US)
Cubic Feet	0.037	Cubic Yards
Cubic Feet per Second	448.83	Gallons (US) per Minute

COMMON UNIT CONVERSIONS *(cont.)*		
Convert From	**Multiply By**	**To Obtain**
Cubic Feet per Second	26,930	Gallons (US) per Hour
Cubic Feet per Minute	0.125	Gallons (US) per Second
Cubic Feet per Minute	448.83	Gallons (US) per Hour
Cubic Feet per Hour	0.002	Gallons (US) per Second
Cubic Feet per Hour	0.125	Gallons (US) per Minute
Cubic Inch	0.016	Liters
Cubic Inch	1.64×10^{-5}	Cubic Meters
Cubic Inch	0.554	Fluid Ounces (US)
Cubic Inch	0.035	Pints (US)
Cubic Inch	0.017	Quarts (US)
Cubic Inch	0.004	Gallons (US)
Cubic Inch	0.001	Cubic Feet
Cubic Inch	2.14×10^{-5}	Cubic Yards
Cubic Meters	1,000	Liters
Cubic Meters	264.172	Gallons (US)
Cubic Meters	35.315	Cubic Feet
Cubic Meters	1.308	Cubic Yards
Cubic Yards	764.555	Liters
Cubic Yards	0.765	Cubic Meters
Cubic Yards	46,656	Cubic Inches

COMMON UNIT CONVERSIONS *(cont.)*		
Convert From	**Multiply By**	**To Obtain**
Cubic Yards	201.974	Gallons (US)
Cubic Yards	27	Cubic Feet
Degrees(angle)	0.01745	Radians
Degrees(angle)	0.00278	Circles
Degrees(angle)	60	Minutes
Feet	304.8	Millimeters
Feet	30.48	Centimeters
Feet	0.305	Meters
Feet	3.05×10^{-4}	Kilometers
Feet	12	Inches
Feet	0.333	Yards
Feet	1.89×10^{-4}	Miles (statute)
Feet	1.65×10^{-4}	Miles (nautical)
Feet of Air	0.0009	Feet of Mercury
Feet of Air	0.00122	Feet of Water
Feet of Air	0.00108	Inches of Mercury
Feet of Air	0.00053	Pounds per Square Inch
Feet of Mercury	30.48	Centimeters of Mercury
Feet of Mercury	13.6086	Feet of Water
Feet of Mercury	163.3	Inches of Water
Feet of Mercury	5.8938	Pounds per Square Inch
Feet of Water	0.0295	Atmospheres

COMMON UNIT CONVERSIONS *(cont.)*

Convert From	Multiply By	To Obtain
Feet of Water	2.2419	Centimeters of Mercury
Feet of Water	0.8826	Inches of Mercury
Feet of Water	304.78	Kilograms per Square Meter
Feet of Water	2988.9	Pascals
Feet of Water	62.424	Pounds per Square Foot
Feet of Water	0.4335	Pounds per Square Inch
Feet per Second	0.305	Meters per Second
Feet per Second	1.097	Kilometers per Hour
Feet per Second	0.592	Knots
Feet per Second	0.682	Miles (statute) per Hour
Foot-Pounds	0.00129	British Thermal Units
Foot-Pounds	1.356	Joules
Foot-Pounds	0.324	Calories
Foot-Pounds	3.766×10^{-7}	Kilowatt-Hours
Gallons (US)	3.785	Liters
Gallons (US)	0.00379	Cubic Meters
Gallons (US)	231	Cubic Inches
Gallons (US)	128	Fluid Ounces (US)
Gallons (US)	8	Pints (US)
Gallons (US)	4	Quarts (US)

COMMON UNIT CONVERSIONS *(cont.)*

Convert From	Multiply By	To Obtain
Gallons (US)	0.134	Cubic Feet
Gallons (US)	4.95×10^{-3}	Cubic Yards
Gallons (US) per Second	8.021	Cubic Feet per Minute
Gallons (US) per Second	481.25	Cubic Feet per Hour
Gallons (US) per Minute	2.23×10^{-3}	Cubic Feet per Second
Gallons (US) per Minute	8.021	Cubic Feet per Hour
Grams	1,000	Milligrams
Grams	0.001	Kilograms
Grams	0.0353	Ounces
Grams	2.20×10^{-3}	Pounds
Hectares	10,000	Square Meters
Hectares	0.01	Square Kilometers
Hectares	107,639	Square Feet
Hectares	11,960	Square Yards
Hectares	2.471	Acres
Hectares	3.86×10^{-3}	Square Miles
Horsepower (US)	42.375	BTU per Minute
Horsepower (US)	2,543	BTU per Hour
Horsepower (US)	550	Foot-Pounds per Second

COMMON UNIT CONVERSIONS (cont.)

Convert From	Multiply By	To Obtain
Horsepower (US)	33,000	Foot-Pounds per Minute
Horsepower (US)	1.014	Horsepower (Metric)
Horsepower (US)	0.7457	Kilowatts
Horsepower (US)	745.7	Watts
Inches	25.4	Millimeters
Inches	2.54	Centimeters
Inches	0.0254	Meters
Inches	0.0833	Feet
Inches	0.0278	Yards
Inches of Mercury	0.0334	Atmospheres
Inches of Mercury	1.133	Feet of Water
Inches of Mercury	3,386	Pascals
Inches of Mercury	70.526	Pounds per Square Foot
Inches of Mercury	0.4912	Pounds per Square Inch
Inches of Water	2.46×10^{-3}	Atmospheres
Inches of Water	0.07355	Inches of Mercury
Inches of Water	25.398	Kilograms per Square Meter
Inches of Water	5.202	Pounds per Square Foot
Inches of Water	0.036	Pounds per Square Inch

COMMON UNIT CONVERSIONS *(cont.)*		
Convert From	**Multiply By**	**To Obtain**
Joules	9.5×10^{-4}	British Thermal Units
Joules	0.239	Calories
Joules	2.778×10^{-7}	Kilowatt-Hours
Joules	0.738	Foot-Pounds
Kilograms	1,000	Grams
Kilograms	35.274	Ounces
Kilograms	2.205	Pounds
Kilograms	1.1×10^{-3}	Short Tons
Kilograms	9.8×10^{-4}	Long Tons
Kilogram-calories	3.968	British Thermal Units
Kilograms per Square Meter	3.28×10^{-3}	Feet of Water
Kilograms per Square Meter	0.2048	Pounds per Square Foot
Kilograms per Square Meter	1.42×10^{-3}	Pounds per Square Inch
Kilometers	1,000	Meters
Kilometers	3,281	Feet
Kilometers	1,094	Yards
Kilometers	0.621	Miles (statute)
Kilometers per Hour	0.278	Meters per Second
Kilometers per Hour	0.54	Knots
Kilometers per Hour	0.911	Feet per Second
Kilometers per Hour	0.621	Miles (statute) per Hour

COMMON UNIT CONVERSIONS *(cont.)*		
Convert From	**Multiply By**	**To Obtain**
Kilopascals	9.87×10^{-3}	Atmospheres
Kilopascals	0.2952	Inches of Mercury
Kilopascals	4.021	Inches of Water
Kilopascals	1,000	Pascals
Kilopascals	20.885	Pounds per Square Foot
Kilopascals	0.145	Pounds per Square Inch
Kilowatts	56.8725	BTU per Minute
Kilowatts	1.341	Horsepower
Kilowatt-Hours	3,412	British Thermal Units
Kilowatt-Hours	3,600,000	Joules
Kilowatt-Hours	860,421	Calories
Kilowatt-Hours	2,655,000	Foot-Pounds
Knots	0.514	Meters per Second
Knots	1.852	Kilometers per Hour
Knots	1.688	Feet per Second
Knots	1.151	Miles (statute) per Hour
Liters	0.001	Cubic Meters
Liters	61.024	Cubic Inches
Liters	33.814	Fluid Ounces (US)
Liters	0.264	Gallons (US)
Liters	0.0353	Cubic Feet
Liters	1.31×10^{-3}	Cubic Yards

COMMON UNIT CONVERSIONS *(cont.)*		
Convert From	**Multiply By**	**To Obtain**
Megapascals	145	Pounds per Square Inch
Meters	1,000	Millimeters
Meters	100	Centimeters
Meters	0.001	Kilometers
Meters	39.37	Inches
Meters	3.281	Feet
Meters	1.094	Yards
Meters	6.21×10^{-4}	Miles (statute)
Meters per Second	3.6	Kilometers per Hour
Meters per Second	1.944	Knots
Meters per Second	3.281	Feet per Second
Meters per Second	2.237	Miles (statute) per Hour
Miles (nautical)	1.1516	Miles (statute)
Miles (statute)	1,609	Meters
Miles (statute)	1.609	Kilometers
Miles (statute)	5,280	Feet
Miles (statute)	1,760	Yards
Miles (statute)	0.8684	Miles (nautical)
Miles (statute) per Hour	0.447	Meters per Second
Miles (statute) per Hour	1.609	Kilometers per Hour

Convert From	Multiply By	To Obtain
Miles (statute) per Hour	0.869	Knots
Miles (statute) per Hour	1.467	Feet per Second
Millimeters	0.1	Centimeters
Millimeters	0.001	Meters
Millimeters	0.0394	Inches
Millimeters	3.28×10^{-3}	Feet
Ounces	28,350	Milligrams
Ounces	28.35	Grams
Ounces	0.0283	Kilograms
Ounces	0.0625	Pounds
Ounces (Fluid, US)	0.0296	Liters
Ounces (Fluid, US)	1.805	Cubic Inches
Ounces (Fluid, US)	7.81×10^{-3}	Gallons (US)
Ounces (Fluid, US)	1.04×10^{-3}	Cubic Feet
Pascals	9.87×10^{-6}	Atmospheres
Pascals	3.35×10^{-4}	Feet of Water
Pascals	2.95×10^{-4}	Inches of Mercury
Pascals	4.01×10^{-3}	Atmospheres
Pascals	0.102	Kilograms per Square Meter
Pascals	0.021	Pounds per Square Foot

COMMON UNIT CONVERSIONS (cont.)

Convert From	Multiply By	To Obtain
Pascals	1.45×10^{-4}	Pounds per Square Inch
Pounds	453.592	Grams
Pounds	0.454	Kilograms
Pounds	4.54×10^{-4}	Tons (Metric)
Pounds	16	Ounces
Pounds	0.0005	Tons (Short)
Pounds	4.46×10^{-4}	Tons (Long)
Pounds per Cubic Foot	16.02	Kilograms per Cubic Meter
Pounds per Cubic Foot	5.79×10^{-4}	Pounds per Cubic Inch
Pounds per Cubic Inch	27.68	Grams per Cubic Centimeter
Pounds per Cubic Inch	27,680	Kilograms per Cubic Meter
Pounds per Cubic Inch	1,728	Pounds per Cubic Foot
Pounds per Square Foot	0.016	Feet of Water
Pounds per Square Foot	4.89	Kilograms per Square Meter
Pounds per Square Foot	0.007	Pounds per Square Inch
Pounds per Square Foot	47.88	Pascals

COMMON UNIT CONVERSIONS (cont.)

Convert From	Multiply By	To Obtain
Pounds per Square Foot	0.048	Kilopascals
Pounds per Square Inch	2.31	Feet of Water
Pounds per Square Inch	27.73	Inches of Water
Pounds per Square Inch	0.0703	Kilograms per Square Centimeter
Pounds per Square Inch	2.036	Inches of Mercury
Pounds per Square Inch	144	Pounds per Square Foot
Pounds per Square Inch	6,895	Pascals
Pounds per Square Inch	6.895	Kilopascals
Pounds per Square Inch	0.0069	Megapascals
Radians	57.3	Degrees
Square Centimeters	100	Square Millimeters
Square Centimeters	0.0001	Square Meters
Square Centimeters	0.155	Square Inches
Square Centimeters	1.08×10^{-3}	Square Feet
Square Centimeters	1.2×10^{-4}	Square Yards
Square Feet	0.0929	Square Meters
Square Feet	9.3×10^{-6}	Hectares

COMMON UNIT CONVERSIONS *(cont.)*

Convert From	Multiply By	To Obtain
Square Feet	9.3×10^{-8}	Square Kilometers
Square Feet	144	Square Inches
Square Feet	0.111	Square Yards
Square Feet	2.3×10^{-5}	Acres
Square Inches	645.16	Square Millimeters
Square Inches	6.452	Square Centimeters
Square Inches	6.45×10^{-4}	Square Meters
Square Inches	6.94×10^{-3}	Square Feet
Square Inches	7.72×10^{-4}	Square Yards
Square Kilometers	1,000,000	Square Meters
Square Kilometers	100	Hectare
Square Kilometers	10,760,000	Square Feet
Square Kilometers	1,196,000	Square Yards
Square Kilometers	247.105	Acres
Square Kilometers	0.386	Square Miles
Square Meters	10,000	Square Centimeters
Square Meters	0.0001	Hectares
Square Meters	0.000001	Square Kilometers
Square Meters	1,550	Square Inches
Square Meters	10.764	Square Feet
Square Meters	1.196	Square Yards
Square Meters	2.47×10^{-4}	Acres
Square Miles	2,590,000	Square Meters
Square Miles	259	Hectares

COMMON UNIT CONVERSIONS *(cont.)*

Convert From	Multiply By	To Obtain
Square Miles	2.59	Square Kilometers
Square Miles	27,880,000	Square Feet
Square Miles	3,098,000	Square Yards
Square Miles	640	Acres
Square Millimeters	0.01	Square Centimeters
Square Millimeters	0.000001	Square Meters
Square Millimeters	1.55×10^{-3}	Square Inches
Square Millimeters	1.08×10^{-5}	Square Feet
Square Yards	0.836	Square Meters
Square Yards	8.36×10^{-5}	Hectares
Square Yards	8.36×10^{-7}	Square Kilometers
Square Yards	1296	Square Inches
Square Yards	9	Square Feet
Square Yards	2.07×10^{-4}	Acres
Tons (Metric)	1,000	Kilograms
Tons (Metric)	2,205	Pounds
Tons (Metric)	1.102	Tons (Short)
Tons (Metric)	0.984	Tons (Long)
Tons (Short)	907.184	Kilograms
Tons (Short)	0.907	Tons (Metric)
Tons (Short)	2000	Pounds
Tons (Short)	0.893	Tons (Long)
Tons (Long)	1,016	Kilograms
Tons (Long)	1.016	Tons (Metric)

COMMON UNIT CONVERSIONS *(cont.)*		
Convert From	**Multiply By**	**To Obtain**
Tons (Long)	2,240	Pounds
Tons (Long)	1.12	Tons (Short)
Watts	3.4121	BTU per Hour
Watts	0.0568	BTU per Minute
Watts	0.0013	Horsepower
Watts	14.34	Calories per Minute
Watt-hours	3.4144	British Thermal Units
Yards	914.4	Millimeters
Yards	91.44	Centimeters
Yards	0.914	Meters
Yards	9.14×10^{-4}	Kilometers
Yards	36	Inches
Yards	3	Feet
Yards	5.68×10^{-4}	Miles (statute)

CONVERTING INCHES TO DECIMALS

Inches	Inches in Decimals	Feet In Decimals	Millimeters	Meters
1/16	0.0625	0.0052	1.5875	0.0016
1/8	0.1250	0.0104	3.1750	0.0032
3/16	0.1875	0.0156	4.7625	0.0048
1/4	0.2500	0.0208	6.3500	0.0064
5/16	0.3125	0.0260	7.9375	0.0079
3/8	0.3750	0.0313	9.5250	0.0095
7/16	0.4375	0.0365	11.1125	0.0111
1/2	0.5000	0.0417	12.7000	0.0127
9/16	0.5625	0.0469	14.2875	0.0143
5/8	0.6250	0.0521	15.8750	0.0159
11/16	0.6875	0.0573	17.4625	0.0175
3/4	0.7500	0.0625	19.0500	0.0191
13/16	0.8125	0.0677	20.6375	0.0206
7/8	0.8750	0.0729	22.2250	0.0222
15/16	0.9375	0.0781	23.8125	0.0238
1	1.0000	0.0833	25.4000	0.0254
2	2.0000	0.1667	50.8000	0.0508
3	3.0000	0.2500	76.2000	0.0762
4	4.0000	0.3333	101.6000	0.1016
5	5.0000	0.4167	127.0000	0.1270
6	6.0000	0.5000	152.4000	0.1524
7	7.0000	0.5833	177.8000	0.1778
8	8.0000	0.6667	203.2000	0.2032
9	9.0000	0.7500	228.6000	0.2286
10	10.0000	0.8333	254.0000	0.2540
11	11.0000	0.9167	279.4000	0.2794
12	12.0000	1.0000	304.8000	0.3048

CONVERTING FAHRENHEIT TEMPERATURE TO CELSIUS TEMPERATURE

Math: Degrees in Celsius = (Degrees in Fahrenheit − 32) × 5/9

Fahrenheit	Celsius	Fahrenheit	Celsius
-30	-34.4	-7	-21.7
-29	-33.9	-6	-21.1
-28	-33.3	-5	-20.6
-27	-32.8	-4	-20
-26	-32.2	-3	-19.4
-25	-31.7	-2	-18.9
-24	-31.1	-1	-18.3
-23	-30.6	0	-17.8
-22	-30	1	-17.2
-21	-29.4	2	-16.7
-20	-28.9	3	-16.1
-19	-28.3	4	-15.6
-18	-27.8	5	-15
-17	-27.2	6	-14.4
-16	-26.7	7	-13.9
-15	-26.1	8	-13.3
-14	-25.6	9	-12.8
-13	-25	10	-12.2
-12	-24.4	11	-11.7
-11	-23.9	12	-11.1
-10	-23.3	13	-10.6
-9	-22.8	14	-10
-8	-22.2	15	-9.4

CONVERTING FAHRENHEIT TEMPERATURE TO CELSIUS TEMPERATURE (cont.)

Math: Degrees in Celsius = (Degrees in Fahrenheit − 32) × 5/9

Fahrenheit	Celsius	Fahrenheit	Celsius
16	-8.9	39	3.9
17	-8.3	40	4.4
18	-7.8	41	5
19	-7.2	42	5.6
20	-6.7	43	6.1
21	-6.1	44	6.7
22	-5.6	45	7.2
23	-5	46	7.8
24	-4.4	47	8.3
25	-3.9	48	8.9
26	-3.3	49	9.4
27	-2.8	50	10
28	-2.2	51	10.6
29	-1.7	52	11.1
30	-1.1	53	11.7
31	-0.6	54	12.2
32	0	55	12.8
33	0.6	56	13.3
34	1.1	57	13.9
35	1.7	58	14.4
36	2.2	59	15
37	2.8	60	15.6
38	3.3	61	16.1

CONVERTING FAHRENHEIT TEMPERATURE TO CELSIUS TEMPERATURE *(cont.)*

Math: Degrees in Celsius = (Degrees in Fahrenheit – 32) × 5/9

Fahrenheit	Celsius	Fahrenheit	Celsius
62	16.7	85	29.4
63	17.2	86	30
64	17.8	87	30.6
65	18.3	88	31.1
66	18.9	89	31.7
67	19.4	90	32.2
68	20	91	32.8
69	20.6	92	33.3
70	21.1	93	33.9
71	21.7	94	34.4
72	22.2	95	35
73	22.8	96	35.6
74	23.3	97	36.1
75	23.9	98	36.7
76	24.4	99	37.2
77	25	100	37.8
78	25.6	101	38.3
79	26.1	102	38.9
80	26.7	103	39.4
81	27.2	104	40
82	27.8	105	40.6
83	28.3	106	41.1
84	28.9	107	41.7

CONVERTING FAHRENHEIT TEMPERATURE TO CELSIUS TEMPERATURE (cont.)

Math: Degrees in Celsius = (Degrees in Fahrenheit – 32) × 5/9

Fahrenheit	Celsius	Fahrenheit	Celsius
108	42.2	131	55
109	42.8	132	55.6
110	43.3	133	56.1
111	43.9	134	56.7
112	44.4	135	57.2
113	45	136	57.8
114	45.6	137	58.3
115	46.1	138	58.9
116	46.7	139	59.4
117	47.2	140	60
118	47.8	141	60.6
119	48.3	142	61.1
120	48.9	143	61.7
121	49.4	144	62.2
122	50	145	62.8
123	50.6	146	63.3
124	51.1	147	63.9
125	51.7	148	64.4
126	52.2	149	65
127	52.8	150	65.6
128	53.3	151	66.1
129	53.9	152	66.7
130	54.4	153	67.2

Math: Degrees in Celsius = (Degrees in Fahrenheit − 32) × 5/9

Fahrenheit	Celsius	Fahrenheit	Celsius
154	67.8	177	80.6
155	68.3	178	81.1
156	68.9	179	81.7
157	69.4	180	82.2
158	70	181	82.8
159	70.6	182	83.3
160	71.1	183	83.9
161	71.7	184	84.4
162	72.2	185	85
163	72.8	186	85.6
164	73.3	187	86.1
165	73.9	188	86.7
166	74.4	189	87.2
167	75	190	87.8
168	75.6	191	88.3
169	76.1	192	88.9
170	76.7	193	89.4
171	77.2	194	90
172	77.8	195	90.6
173	78.3	196	91.1
174	78.9	197	91.7
175	79.4	198	92.2
176	80	199	92.8

CONVERTING FAHRENHEIT TEMPERATURE TO CELSIUS TEMPERATURE *(cont.)*

Math: Degrees in Celsius = (Degrees in Fahrenheit − 32) × 5/9

Fahrenheit	Celsius	Fahrenheit	Celsius
200	93.3	221	105
201	93.9	222	105.6
202	94.4	223	106.1
203	95	224	106.7
204	95.6	225	107.2
205	96.1	226	107.8
206	96.7	227	108.3
207	97.2	228	108.9
208	97.8	229	109.4
209	98.3	230	110
210	98.9	231	110.6
211	99.4	232	111.1
212	100	233	111.7
213	100.6	234	112.2
214	101.1	235	112.8
215	101.7	236	113.3
216	102.2	237	113.9
217	102.8	238	114.4
218	103.3	239	115
219	103.9	240	115.6
220	104.4	241	116.1

CONVERTING CELSIUS TEMPERATURE TO FAHRENHEIT TEMPERATURE

Math: Degrees in Fahrenheit = Degrees in Celsius $\times$ 9/5 + 32

Celsius	Fahrenheit	Celsius	Fahrenheit
-35	-31	-9	15.8
-34	-29.2	-8	17.6
-33	-27.4	-7	19.4
-32	-25.6	-6	21.2
-31	-23.8	-5	23
-30	-22	-4	24.8
-29	-20.2	-3	26.6
-28	-18.4	-2	28.4
-27	-16.6	-1	30.2
-26	-14.8	0	32
-25	-13	1	33.8
-24	-11.2	2	35.6
-23	-9.4	3	37.4
-22	-7.6	4	39.2
-21	-5.8	5	41
-20	-4	6	42.8
-19	-2.2	7	44.6
-18	-0.4	8	46.4
-17	1.4	9	48.2
-16	3.2	10	50
-15	5	11	51.8
-14	6.8	12	53.6
-13	8.6	13	55.4
-12	10.4	14	57.2
-11	12.2	15	59
-10	14	16	60.8

CONVERTING CELSIUS TEMPERATURE
TO FAHRENHEIT TEMPERATURE *(cont.)*

Math: Degrees in Fahrenheit = Degrees in Celsius × 9/5 + 32

Celsius	Fahrenheit	Celsius	Fahrenheit
17	62.6	43	109.4
18	64.4	44	111.2
19	66.2	45	113
20	68	46	114.8
21	69.8	47	116.6
22	71.6	48	118.4
23	73.4	49	120.2
24	75.2	50	122
25	77	51	123.8
26	78.8	52	125.6
27	80.6	53	127.4
28	82.4	54	129.2
29	84.2	55	131
30	86	56	132.8
31	87.8	57	134.6
32	89.6	58	136.4
33	91.4	59	138.2
34	93.2	60	140
35	95	61	141.8
36	96.8	62	143.6
37	98.6	63	145.4
38	100.4	64	147.2
39	102.2	65	149
40	104	66	150.8
41	105.8	67	152.6
42	107.6	68	154.4

CONVERTING CELSIUS TEMPERATURE
TO FAHRENHEIT TEMPERATURE *(cont.)*

Math: Degrees in Fahrenheit = Degrees in Celsius $\times$ 9/5 + 32

Celsius	Fahrenheit	Celsius	Fahrenheit
69	156.2	95	203
70	158	96	204.8
71	159.8	97	206.6
72	161.6	98	208.4
73	163.4	99	210.2
74	165.2	100	212
75	167	101	213.8
76	168.8	102	215.6
77	170.6	103	217.4
78	172.4	104	219.2
79	174.2	105	221
80	176	106	222.8
81	177.8	107	224.6
82	179.6	108	226.4
83	181.4	109	228.2
84	183.2	110	230
85	185	111	231.8
86	186.8	112	233.6
87	188.6	113	235.4
88	190.4	114	237.2
89	192.2	115	239
90	194	116	240.8
91	195.8	117	242.6
92	197.6	118	244.4
93	199.4	119	246.2
94	201.2	120	248

CONVERTING CELSIUS TEMPERATURE
TO FAHRENHEIT TEMPERATURE *(cont.)*

Math: Degrees in Fahrenheit = Degrees in Celsius $\times$ 9/5 + 32

Celsius	Fahrenheit	Celsius	Fahrenheit
121	249.8	145	293
122	251.6	146	294.8
123	253.4	147	296.6
124	255.2	148	298.4
125	257	149	300.2
126	258.8	150	302
127	260.6	151	303.8
128	262.4	152	305.6
129	264.2	153	307.4
130	266	154	309.2
131	267.8	155	311
132	269.6	156	312.8
133	271.4	157	314.6
134	273.2	158	316.4
135	275	159	318.2
136	276.8	160	320
137	278.6	161	321.8
138	280.4	162	323.6
139	282.2	163	325.4
140	284	164	327.2
141	285.8	165	329
142	287.6	166	330.8
143	289.4	167	332.6
144	291.2	168	334.4

BOILER CONVERSION RATIOS

From	Multiply	To Obtain
sq. ft. EDR steam	240	Btu per hr.
sq. ft. EDR water	160	Btu per hr.
Boiler hp	33.5	MBtu per hr.
Boiler hp	140	sq. ft. EDR steam
Boiler hp	223	sq. ft. EDR water
Boiler hp	34.5	lb. per hr. steam
lb. per hr. steam	970	Btu per hr.
sq. ft. EDR steam	0.247	lb. per hr. steam
sq. ft. EDR water	0.155	lb. per hr. steam
To obtain above	Divide by above	Starting with above

CONVERTING BOILER HORSE POWER TO BTU PER HOUR

Math: BTU per Hour = Boiler Horse Power (BHP) $\times$ 33,479
Or: Thousand BTU per Hour (MBH) = Boiler Horse Power (BHP) $\times$ 33,479

BHP	MBH	BHP	MBH
4.5	151	125	4,185
6.0	201	150	5,022
9.0	301	200	6,696
15.0	502	250	8,370
20.0	670	300	10,044
30.0	1,004	350	11,718
45.0	1,507	400	13,392
60.0	2,009	500	16,740
30.0	1,004	600	20,087
40.0	1,339	700	23,435
50.0	1,674	750	25,109
60.0	2,009	800	26,783
70.0	2,344	850	28,457
80.0	2,678	950	31,805
100	3,348	1000	33,479

BTU CONTENT OF FUEL OIL

Grade or Type	Unit	BTU
No.1 Oil	Gallon	137,400
No.2 Oil	Gallon	139,600
No.3 Oil	Gallon	141,800
No.4 Oil	Gallon	145,100
No.5 Oil	Gallon	148,800
No.6 Oil	Gallon	152,400
Natural Gas	Cubic ft.	950 to 1150
Propane	Cubic ft.	2550
Butane	Cubic ft.	3200

PROPERTIES OF LP GAS

Property	Butane	Propane
Btu per cu. ft., 60° F	3,280	2,516
Btu per lb.	21,221	21,591
Btu per gal.	102,032	91,547
Cubic ft. per lb.	6.506	8.58
Cubic ft. per gal.	31.26	36.69
Lbs. per gal.	4.81	4.24

CONVERTING BTU TO TONS OF REFRIGERATION

BTU per hour	Tons of Refrigeration	BTU per hour	Tons of Refrigeration
6,000	½ ton	66,000	5½ ton
12,000	2 ton	72,000	6 ton
18,000	1½ ton	78,000	6½ ton
24,000	2 ton	84,000	7 ton
30,000	2½ ton	90,000	7½ ton
36,000	3 ton	96,000	8 ton
42,000	3½ ton	102,000	8½ ton
48,000	4 ton	108,000	9 ton
54,000	4½ ton	114,000	9½ ton
60,000	5 ton	120,000	10 ton

ESTIMATING REFRIGERANT CHARGE IN SYSTEMS

System	Lbs. of Refrigerant per Ton
Package units, self-contained	3.0
Package units, split systems	3.5
Built-up, close-coupled	4.0
Built-up, remote	6.0

FRESH AIR REQUIREMENTS FOR DIFFERENT PROJECTS

The numbers listed below are average values. Please verify with specific project conditions as well as the local building codes.

Project Type	Air Changes per Hour
Auditorium	6
Attic Space	12 to 15
Bakeries	20
Boiler Room	15 to 20
Bowling Alleys	12
Churches	6 to 8
Club Room	12
Dining Room	5 to 12
Factories	10 to 15
Foundries	15 to 20
Galvanizing Plants	20 to 30
Garages	12 to 30
Homes	9 to 17
Kitchens	20 to 30
Laundries	15 to 20
Mills	15 to 20
Offices	4 to 10
Restaurants	8 to 12
Shops	15 to 20

ATTIC VENT SIZING GUIDE

The data listed below are the **minimum** required for attic floor area given. Please verify with specific project conditions as well as the local building codes.

Attic Floor Area	Turbine Ventilators	Inlet Louver Area	Eave Vents	Gable Vents
1200 sq. ft.	2 ea., 12"	4 sq. ft.	6 ea. 8" × 16"	2 ea. 14" × 24"
1500 sq. ft.	2 ea., 14"	5 sq. ft.	6 ea. 8" × 16"	2 ea. 14" × 24"
1800 sq. ft.	3 ea., 12"	6 sq. ft.	8 ea. 8" × 16"	4 ea. 12" × 18"
2100 sq. ft.	3 ea., 14"	7 sq. ft.	8 ea. 8" × 16"	2 ea. 12" × 18" & 14" × 24"
2400 sq. ft.	4 ea., 12"	8 sq. ft.	10 ea. 8" × 16"	2 ea. 12" × 18" & 14" × 24"

WATER COIL PIPING SIZING GUIDE
Up to 16,000 CFM

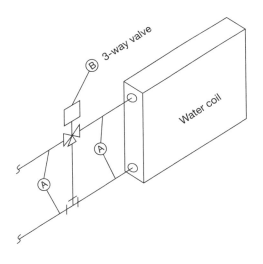

A	B
1"	¾"
1¼"	1"
1½"	1¼"
2"	1½"
2½"	2"
3"	2½"

7-93

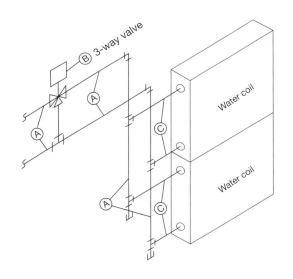

A	B	C
2½"	2"	2"
3"	2½"	2½"
4"	3"	3"

WATER COIL PIPING SIZING GUIDE *(cont.)*

Above 26,000 CFM

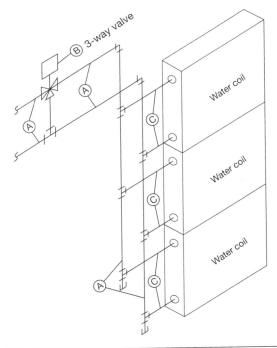

A	B	C
4"	3"	2½"
6"	4"	3"
8"	6"	4"
10"	8"	6"

CONVERTING ROUND DUCT AREAS TO SQUARE FEET

Duct Diameter (inches)	Duct Diameter (mm)	Area (Ft²)	Area (m²)
8	203	0.3491	0.032
10	254	0.5454	0.051
12	305	0.7854	0.073
14	356	1.069	0.099
16	406	1.396	0.13
18	457	1.767	0.29
20	508	2.182	0.203
22	559	2.64	0.245
24	609	3.142	2.292
26	660	3.687	0.342
28	711	4.276	0.397
30	762	4.9	0.455
32	813	5.585	0.519
34	864	6.305	0.586
36	914	7.069	0.657
38	965	7.786	0.732
40	1016	8.727	0.811
42	1067	9.62	0.894
44	1119	10.56	0.981
46	1168	11.54	1.072
48	1219	12.57	1.168
50	1270	13.67	1.27
52	1321	14.75	1.37
54	1372	15.9	1.477
56	1422	17.1	1.586
58	1473	18.35	1.705
60	1524	19.63	1.824

FIBROUS GLASS DUCT STRETCH-OUT AREAS

Note: The table shows material required for rectangular ducts fabricated from 1"-thick fibrous glass board; includes allowance for overlap and 8" grooving.

Width + Depth	Sq. Ft. per Lin. Ft.	Width + Depth	Sq. Ft. per Lin. Ft.
10	2.33	33	6.17
11	2.5	34	6.33
12	2.67	35	6.5
13	2.83	36	6.67
14	3	37	6.83
15	3.17	38	7
16	3.33	39	7.17
17	3.5	40	7.33
18	3.67	41	7.5
19	3.83	42	7.67
20	4	43	7.83
21	4.17	44	8
22	4.33	45	8.17
23	4.5	46	8.33
24	4.67	47	8.5
25	4.83	48	8.67
26	5	49	8.83
27	5.17	50	9
28	5.33	51	9.17
29	5.5	52	9.33
30	5.67	53	9.5
31	5.83	54	9.67
32	6	55	9.83

FIBROUS GLASS DUCT STRETCH-OUT AREAS *(cont.)*

Note: The table shows material required for rectangular ducts fabricated from 1"-thick fibrous glass board; includes allowance for overlap and 8" grooving.

Width + Depth	Sq. Ft. per Lin. Ft.	Width + Depth	Sq. Ft. per Lin. Ft.
56	10	79	13.83
57	10.17	80	14
58	10.33	81	14.17
59	10.5	82	14.33
60	10.67	83	14.5
61	10.83	84	14.67
62	11	85	14.83
63	11.17	86	15
64	11.33	87	15.17
65	11.5	88	15.33
66	11.67	89	15.5
67	11.83	90	15.67
68	12	91	15.83
69	12.17	92	16
70	12.33	93	16.17
71	12.5	94	16.33
72	12.67	95	16.5
73	12.83	96	16.67
74	13	97	16.83
75	13.17	98	17
76	13.33	99	17.17
77	13.5	100	17.33
78	13.67		

WEIGHTS OF STEEL SHEET DUCT MATERIALS

Galvanized Steel

U.S. Gauge	Decimals Equivalent (Inches)	Lbs. per Sq. Ft.	Lbs. per Sheet		
			36" × 96"	48" × 96"	48" × 120"
28	0.020	0.781	18.75	N/A	N/A
26	0.022	0.906	21.75	29.0	36.2
24	0.028	1.156	27.75	37.0	46.2
22	0.034	1.406	33.75	45.0	56.2
20	0.040	1.656	39.75	53.0	66.2
18	0.052	2.156	51.75	70.0	86.2
16	0.064	2.656	63.75	85.0	102.2
14	0.080	3.281	78.75	105.0	131.2
12	0.112	4.531	108.75	145.0	181.2
10	0.142	5.781	138.75	185.0	231.2

WEIGHTS OF STEEL SHEET DUCT MATERIALS (cont.)

Hot-Rolled Steel

U.S. Gauge	Decimals Equivalent (Inches)	Lbs. per Sq. Ft.	Lbs. per Sheet		
			36" × 96"	48" × 96"	48" × 120"
26	0.018	0.750	18.0	24.0	30.0
24	0.024	1.000	24.0	32.0	40.0
22	0.030	1.250	30.0	40.0	50.0
20	0.036	1.500	36.0	48.0	60.0
18	0.048	2.000	48.0	64.0	80.0
16	0.057	2.500	60.0	80.0	100.0
14	0.075	3.125	75.0	100.0	125.0
12	0.108	4.250	102.0	138.0	170.0
10	0.135	5.625	137.0	180.0	225.0

WEIGHTS OF STEEL SHEET DUCT MATERIALS (cont.)

Stainless Steel

U.S. Gauge	Decimals Equivalent (Inches)	Lbs. per Sq. Ft.	Lbs. per Sheet		
			36" × 96"	48" × 96"	48" × 120"
28	0.016	0.66	15.8	21.1	26.4
26	0.019	0.79	18.9	25.2	31.6
24	0.025	1.05	25.2	33.6	42.0
22	0.031	1.31	31.5	42.0	52.5
20	0.038	1.58	37.8	50.4	63.0
18	0.050	2.10	50.4	61.2	84.0
16	0.063	2.63	63.0	84.0	105.0
14	0.078	3.28	78.7	104.9	131.2
12	0.109	4.60	110.0	147.0	183.8

WEIGHTS OF NONFERROUS SHEET DUCT MATERIALS

Aluminum 3003

B & S Gauge	Decimals Equivalent (Inches)	Lbs. per Sq. Ft.	Lbs. per Sheet		
			36" × 96"	48" × 96"	48" × 120"
26	0.016	0.226	5.4	7.2	9.0
24	0.020	0.282	6.8	9.0	11.3
22	0.025	0.357	8.6	11.4	14.3
20	0.032	0.450	10.8	14.4	18.0
18	0.040	0.568	13.6	18.7	22.7
16	0.051	0.716	17.2	22.9	28.6
14	0.064	0.903	21.7	28.9	36.1
12	0.071	1.000	24.0	32.0	40.0
10	0.080	1.130	27.1	36.2	45.2

WEIGHTS OF NONFERROUS SHEET DUCT MATERIALS (cont.)

Cold-rolled Copper

B & S Gauge	Decimals Equivalent (Inches)	Lbs. per Sq. Ft.	Lbs. per Sheet		
			36" × 96"	48" × 96"	48" × 120"
24	0.021	1.000	24	32	40
23	0.024	1.125	30	40	50
20	0.032	1.500	36	48	64
18	0.040	1.875	48	64	80
16	0.051	2.375	54	72	90
15	0.053	2.500	60	80	100

7-103

WEIGHTS OF GALVANIZED STEEL BANDS

Band Size	Lbs. per Lin. Ft.
1/8" × 1"	0.425
1/8" × 1 1/4"	0.531
1/8" × 1/2"	0.638
1/8" × 2"	0.850
3/16" × 1"	0.670
3/16" × 1 1/4"	0.837
3/16" × 1/2"	1.000
3/16" × 2"	1.340
1/4" × 1"	0.900
1/4" × 1 1/2"	1.350
1/4" × 2"	1.780

WEIGHTS OF METAL ANGLES
(EQUAL LEGS, LBS. PER LIN. FT.)

Size	Galvanized Steel	Hot-rolled Steel	Aluminum
$\frac{1}{8}$" × 1"	0.84	0.80	0.28
$\frac{1}{8}$" × 1$\frac{1}{4}$"	1.06	1.01	0.36
$\frac{1}{8}$" × 1$\frac{1}{2}$"	1.29	1.23	0.44
$\frac{1}{8}$" × 1$\frac{3}{4}$"	1.51	1.44	0.51
$\frac{1}{8}$" × 2"	1.73	1.65	0.59
$\frac{3}{16}$" × 1"	1.22	1.16	0.41
$\frac{3}{16}$" × 1$\frac{1}{4}$"	1.55	1.48	0.53
$\frac{3}{16}$" × 1$\frac{1}{2}$"	1.89	1.80	0.64
$\frac{3}{16}$" × 1$\frac{3}{4}$"	2.23	2.12	0.75
$\frac{3}{16}$" × 2"	2.56	2.44	0.87
$\frac{1}{4}$" × 1$\frac{1}{2}$"	2.46	2.34	0.83
$\frac{1}{4}$" × 1$\frac{3}{4}$"	2.91	2.77	0.98
$\frac{1}{4}$" × 2"	3.35	3.19	1.14
$\frac{1}{4}$" × 2$\frac{1}{2}$"	4.26	4.10	1.45
$\frac{1}{4}$" × 3"	5.15	4.90	1.70
$\frac{3}{8}$" × 2"	4.93	4.70	1.65
$\frac{3}{8}$" × 2$\frac{1}{2}$"	6.20	5.90	2.11
$\frac{3}{8}$" × 3"	7.55	7.20	2.55

WEIGHTS OF FLOOR PLATES (SKID-RESISTANT, RAISED PATTERN)

Thickness	Lbs. per Sq. Ft.
1/8"	6.15
3/16"	8.70
1/4"	11.25
5/16"	13.80
3/8"	16.35
7/16"	18.90
1/2"	21.45
5/8"	26.55
3/4"	31.65
7/8"	36.75
1"	41.85

WEIGHTS OF ROUND HANGER ROD (THREADED JOINTS, LBS. PER LIN. FT.)

Size	Steel	Aluminum	Brass
1/4"	0.167	0.06	0.181
5/16"	0.261	0.09	0.283
3/8"	0.376	0.13	0.407
7/16"	0.511	0.18	0.554
1/2"	0.668	0.24	0.723
5/8"	1.043	0.37	1.130
3/4"	1.502	0.54	1.163
7/8"	2.044	0.73	2.220
1"	2.670	0.96	2.900
1 1/4"	4.172	1.50	4.520

WEIGHTS OF GALVANIZED BELT GUARDS

Nominal Wheel Diameter	Pounds
12"	23
15"	25
16"	26
18"	27
20"	28
22"	29
24"	30
28"	31
30"	32
34"	33
36"	34
40"	35
44"	38
48"	45
54"	52
60"	61
66"	70
72"	84
80"	100

PROPERTIES OF STEEL AND PLATE IRON

Standard Gauge	Thickness (in.)	Weight (lbs./sq. ft.)
0000000	$1/2$	20.00
000000	$15/32$	18.75
00000	$7/16$	17.50
0000	$13/32$	16.25
000	$3/8$	15.00
00	$11/32$	13.75
0	$5/16$	12.50
1	$9/32$	11.25
2	$17/64$	10.62
3	$1/4$	10.00
4	$15/64$	9.37
5	$7/32$	8.75
6	$13/64$	8.12
7	$3/16$	7.50
8	$11/64$	6.87
9	$5/32$	6.25
10	$9/64$	5.62
11	$1/8$	5.00
12	$7/64$	4.38
13	$3/32$	3.75
14	$5/64$	3.13
15	$9/128$	2.81
16	$1/16$	2.50

PROPERTIES OF STEEL AND PLATE IRON *(cont.)*

Standard Gauge	Thickness (in.)	Weight (lbs./sq. ft.)
17	$9/160$	2.25
18	$1/20$	2.00
19	$7/160$	1.75
20	$3/80$	1.50
21	$11/320$	1.37
22	$1/32$	1.25
23	$9/320$	1.12
24	$1/40$	1.00
25	$7/320$	0.87
26	$3/160$	0.75
27	$11/640$	0.69
28	$11/64$	0.62
29	$9/640$	0.56
30	$1/80$	0.50
31	$7/640$	0.44
32	$13/1280$	0.41
33	$3/320$	0.37
34	$11/1280$	0.34
35	$5/640$	0.31
36	$9/1280$	0.28
37	$17/2560$	0.27
38	$1/160$	0.25

ESTIMATING GALVANIZED STEEL RECTANGULAR DUCTWORK

Ductwork Gauge	Weight (lb./sq. ft.)	Section Long Side Max Dimension (in.)
26	0.906	12"
24	1.156	30"
22	1.406	54"
20	1.656	84"
18	2.156	85" and up
16	2.656	85" and up

Estimating Example

Find the weight for 34" × 20" galvanized rectangular ductwork 15' long

Calculation:

Section long side dimension is 34", more than 30" but less than 54"

So this is a 22 gauge ductwork

Area = (34 + 20) × 2/12 × 15 = 135 sq. ft.

Add 25% waste for bracing, hangers, waste, and seams

Area = 135 × (1 + 25%) = 169 sq. ft.

Weight = 169 sq. ft. × 1.406 lbs./sq. ft. = 238 lbs.

WEIGHTS AND AREAS OF GALVANIZED STEEL RECTANGULAR DUCTS

Note: Includes 25% waste for bracing, hangers, waste, and seams

Width + Depth (Inches)	Sq. Ft. per Lin. Ft.	Lbs. per Lin. Ft.					
		26 Gauge	24 Gauge	22 Gauge	20 Gauge	18 Gauge	16 Gauge
10	1.67	1.89	2.41	2.93	N/A	N/A	N/A
11	1.83	2.08	2.65	3.22	N/A	N/A	N/A
12	2.00	2.27	2.89	3.52	N/A	N/A	N/A
13	2.17	2.45	3.13	3.81	N/A	N/A	N/A
14	2.33	2.64	3.37	4.10	N/A	N/A	N/A
15	2.50	2.83	3.61	4.39	N/A	N/A	N/A
16	2.67	3.02	3.85	4.69	N/A	N/A	N/A
17	2.83	3.21	4.09	4.98	N/A	N/A	N/A
18	3.00	3.40	4.34	5.27	N/A	N/A	N/A
19	3.17	3.59	4.58	5.57	N/A	N/A	N/A
20	3.33	3.78	4.82	5.86	N/A	N/A	N/A
21	3.50	3.96	5.06	6.15	N/A	N/A	N/A
22	3.67	4.15	5.30	6.44	N/A	N/A	N/A
23	3.83	4.34	5.54	6.74	N/A	N/A	N/A

WEIGHTS AND AREAS OF GALVANIZED STEEL RECTANGULAR DUCTS (cont.)

Note: Includes 25% waste for bracing, hangers, waste, and seams

Width + Depth (Inches)	Sq. Ft. per Lin. Ft.	Lbs. per Lin. Ft.					
		26 Gauge	24 Gauge	22 Gauge	20 Gauge	18 Gauge	16 Gauge
24	4.00	4.53	5.78	7.03	N/A	N/A	N/A
25	4.17	4.72	6.02	7.32	N/A	N/A	N/A
26	4.33	4.91	6.26	7.62	N/A	N/A	N/A
27	4.50	N/A	6.50	7.91	9.32	12.13	N/A
28	4.67	N/A	6.74	8.20	9.66	12.58	N/A
29	4.83	N/A	6.98	8.49	10.01	13.03	N/A
30	5.00	N/A	7.23	8.79	10.35	13.48	N/A
31	5.17	N/A	7.47	9.08	10.70	13.92	N/A
32	5.33	N/A	7.71	9.37	11.04	14.37	N/A
33	5.50	N/A	7.95	9.67	11.39	14.82	N/A
34	5.67	N/A	8.19	9.96	11.73	15.27	N/A
35	5.83	N/A	8.43	10.25	12.08	15.72	N/A
36	6.00	N/A	8.67	10.55	12.42	16.17	N/A
37	6.17	N/A	8.91	10.84	12.77	16.62	N/A

WEIGHTS AND AREAS OF GALVANIZED STEEL RECTANGULAR DUCTS (cont.)

Note: Includes 25% waste for bracing, hangers, waste, and seams

Width + Depth (Inches)	Sq. Ft. per Lin. Ft.	Lbs. per Lin. Ft.						
		26 Gauge	24 Gauge	22 Gauge	20 Gauge	18 Gauge	16 Gauge
38	6.33	N/A	9.15	11.13	13.11	17.07	N/A
39	6.50	N/A	9.39	11.42	13.46	17.52	N/A
40	6.67	N/A	9.63	11.72	13.80	17.97	N/A
41	6.83	N/A	9.87	12.01	14.15	18.42	N/A
42	7.00	N/A	10.12	12.30	14.49	18.87	N/A
43	7.17	N/A	10.36	12.60	14.84	19.31	N/A
44	7.33	N/A	10.60	12.89	15.18	19.76	N/A
45	7.50	N/A	10.84	13.18	15.53	20.21	N/A
46	7.67	N/A	11.08	13.47	15.87	20.66	N/A
47	7.83	N/A	11.32	13.77	16.22	21.11	N/A
48	8.00	N/A	11.56	14.06	16.56	21.56	26.56
49	8.17	N/A	11.80	14.35	16.91	22.01	27.11
50	8.33	N/A	12.04	14.65	17.25	22.46	27.67
51	8.50	N/A	12.28	14.94	17.60	22.91	28.22

WEIGHTS AND AREAS OF GALVANIZED STEEL RECTANGULAR DUCTS (cont.)

Note: Includes 25% waste for bracing, hangers, waste, and seams

Width + Depth (Inches)	Sq. Ft. per Lin. Ft.	Lbs. per Lin. Ft.						
		26 Gauge	24 Gauge	22 Gauge	20 Gauge	18 Gauge	16 Gauge
52	8.67	N/A	12.52	15.23	17.94	23.36	28.77
53	8.83	N/A	12.76	15.52	18.29	23.81	29.33
54	9.00	N/A	13.01	15.82	18.63	24.26	29.88
55	9.17	N/A	13.25	16.11	18.98	24.70	30.43
56	9.33	N/A	13.49	16.40	19.32	25.15	30.99
57	9.50	N/A	13.73	16.70	19.67	25.60	31.54
58	9.67	N/A	13.97	16.99	20.01	26.05	32.09
59	9.83	N/A	14.21	17.28	20.36	26.50	32.65
60	10.00	N/A	14.45	17.58	20.70	26.95	33.20
61	10.17	N/A	N/A	17.87	21.05	27.40	33.75
62	10.33	N/A	N/A	18.16	21.39	27.85	34.31
63	10.50	N/A	N/A	18.45	21.74	28.30	34.86
64	10.67	N/A	N/A	18.75	22.08	28.75	35.41
65	10.83	N/A	N/A	19.04	22.43	29.20	35.97

WEIGHTS AND AREAS OF GALVANIZED STEEL RECTANGULAR DUCTS (cont.)

Note: Includes 25% waste for bracing, hangers, waste, and seams

Width + Depth (Inches)	Sq. Ft. per Lin. Ft.	Lbs. per Lin. Ft.						
		26 Gauge	24 Gauge	22 Gauge	20 Gauge	18 Gauge	16 Gauge
66	11.00	N/A	N/A	19.33	22.77	29.65	36.52
67	11.17	N/A	N/A	19.63	23.12	30.09	37.07
68	11.33	N/A	N/A	19.92	23.46	30.54	37.63
69	11.50	N/A	N/A	20.21	23.81	30.99	38.18
70	11.67	N/A	N/A	20.50	24.15	31.44	38.73
71	11.83	N/A	N/A	20.80	24.50	31.89	39.29
72	12.00	N/A	N/A	21.09	24.84	32.34	39.84
73	12.17	N/A	N/A	21.38	25.19	32.79	40.39
74	12.33	N/A	N/A	21.68	25.53	33.24	40.95
75	12.50	N/A	N/A	21.97	25.88	33.69	41.50
76	12.67	N/A	N/A	22.26	26.22	34.14	42.05
77	12.83	N/A	N/A	22.55	26.57	34.59	42.61
78	13.00	N/A	N/A	22.85	26.91	35.04	43.16
79	13.17	N/A	N/A	23.14	27.26	35.48	43.71

WEIGHTS AND AREAS OF GALVANIZED STEEL RECTANGULAR DUCTS (cont.)

Note: Includes 25% waste for bracing, hangers, waste, and seams

Width + Depth (Inches)	Sq. Ft. per Lin. Ft.	Lbs. per Lin. Ft.						
		26 Gauge	24 Gauge	22 Gauge	20 Gauge	18 Gauge	16 Gauge
80	13.33	N/A	N/A	23.43	27.60	35.93	44.27
81	13.50	N/A	N/A	23.73	27.95	36.38	44.82
82	13.67	N/A	N/A	24.02	28.29	36.83	45.37
83	13.83	N/A	N/A	24.31	28.64	37.28	45.93
84	14.00	N/A	N/A	24.61	28.98	37.73	46.48
85	14.17	N/A	N/A	24.90	29.33	38.18	47.03
86	14.33	N/A	N/A	25.19	29.67	38.63	47.59
87	14.50	N/A	N/A	25.48	30.02	39.08	48.14
88	14.67	N/A	N/A	25.78	30.36	39.53	48.69
89	14.83	N/A	N/A	26.07	30.71	39.98	49.25
90	15.00	N/A	N/A	26.36	31.05	40.43	49.80
91	15.17	N/A	N/A	26.66	31.40	40.87	50.35
92	15.33	N/A	N/A	26.95	31.74	41.32	50.91
93	15.50	N/A	N/A	27.24	32.09	41.77	51.46

WEIGHTS AND AREAS OF GALVANIZED STEEL RECTANGULAR DUCTS *(cont.)*

Note: Includes 25% waste for bracing, hangers, waste, and seams

Width + Depth (Inches)	Sq. Ft. per Lin. Ft.	Lbs. per Lin. Ft.						
		26 Gauge	24 Gauge	22 Gauge	20 Gauge	18 Gauge	16 Gauge
94	15.67	N/A	N/A	27.53	32.43	42.22	52.01
95	15.83	N/A	N/A	27.83	32.78	42.67	52.57
96	16.00	N/A	N/A	28.12	33.12	43.12	53.12
97	16.17	N/A	N/A	28.41	33.47	43.57	53.67
98	16.33	N/A	N/A	28.71	33.81	44.02	54.23
99	16.50	N/A	N/A	29.00	34.16	44.47	54.78
100	16.67	N/A	N/A	29.29	34.50	44.92	55.33
101	16.83	N/A	N/A	29.58	34.85	45.37	55.89
102	17.00	N/A	N/A	29.88	35.19	45.82	56.44
103	17.17	N/A	N/A	30.17	35.54	46.26	56.99
104	17.33	N/A	N/A	30.46	35.88	46.71	57.55
105	17.50	N/A	N/A	30.76	36.23	47.16	58.10
106	17.67	N/A	N/A	31.05	36.57	47.61	58.65
107	17.83	N/A	N/A	31.34	36.92	48.06	59.21

WEIGHTS AND AREAS OF GALVANIZED STEEL RECTANGULAR DUCTS (cont.)

Note: Includes 25% waste for bracing, hangers, waste, and seams

Width + Depth (Inches)	Sq. Ft. per Lin. Ft.	Lbs. per Lin. Ft.						
		26 Gauge	24 Gauge	22 Gauge	20 Gauge	18 Gauge	16 Gauge
108	18.00	N/A	N/A	31.64	37.26	48.51	59.76
109	18.17	N/A	N/A	31.93	37.61	48.96	60.31
110	18.33	N/A	N/A	32.22	37.95	49.41	60.87
111	18.50	N/A	N/A	32.51	38.30	49.86	61.42
112	18.67	N/A	N/A	32.81	38.64	50.31	61.97
113	18.83	N/A	N/A	33.10	38.99	50.76	62.53
114	19.00	N/A	N/A	33.39	39.33	51.21	63.08
115	19.17	N/A	N/A	33.69	39.68	51.65	63.63
116	19.33	N/A	N/A	33.98	40.02	52.10	64.19
117	19.50	N/A	N/A	34.27	40.37	52.55	64.74
118	19.67	N/A	N/A	34.56	40.71	53.00	65.29
119	19.83	N/A	N/A	34.86	41.06	53.45	65.85
120	20.00	N/A	N/A	35.15	41.40	53.90	66.40

WEIGHTS OF GALVANIZED STEEL RECTANGULAR ELBOWS

Pounds per Elbow

Size	4"	6"	8"	10"	12"	14"	15"	18"	20"
4"	1.05	1.45	1.95	2.60	3.30	5.30	6.50	7.80	9.25
6"	1.45	1.95	2.60	3.30	3.70	5.80	7.75	8.75	9.75
8"	1.95	2.60	3.30	3.70	4.25	6.50	8.50	9.75	10.40
10"	2.60	3.30	3.70	4.25	4.85	7.30	9.50	10.70	11.20
12"	3.30	3.70	4.25	4.85	5.60	8.20	10.50	11.70	12.20
14"	5.30	5.80	6.50	7.30	8.20	9.25	11.50	12.60	13.20
16"	6.50	7.50	8.50	9.50	10.50	11.50	12.50	13.40	14.40
18"	7.80	8.75	9.75	10.70	11.70	12.60	13.40	14.50	15.70
20"	9.25	9.75	10.40	11.20	12.20	13.80	14.40	15.70	17.20
22"	10.80	11.80	12.80	13.80	14.80	25.80	16.80	17.80	18.80
24"	12.60	13.60	14.60	15.60	16.60	17.60	18.60	19.60	20.50
26"	14.40	15.40	16.40	17.40	18.40	19.40	20.40	21.40	22.30

WEIGHTS OF GALVANIZED STEEL RECTANGULAR ELBOWS (cont.)

Pounds per Elbow

Size	4"	6"	8"	10"	12"	14"	15"	18"	20"
28"	16.40	17.40	18.40	19.40	20.40	21.40	22.40	23.40	24.30
30"	18.50	19.50	20.50	21.50	22.50	23.50	24.50	25.50	26.40
32"	24.30	25.70	27.00	28.40	29.70	31.10	32.40	33.70	35.00
34"	28.20	29.40	30.60	31.80	33.00	34.30	35.50	36.70	37.90
36"	31.30	32.50	33.70	34.90	36.10	37.30	38.50	39.70	40.90
38"	34.50	35.70	36.90	38.10	39.30	40.50	41.70	42.90	44.10
40"	37.90	39.10	40.30	41.50	42.70	43.90	45.10	46.30	47.50
42"	41.40	42.60	43.80	45.00	46.20	47.40	48.60	49.80	51.00
44"	45.10	46.40	47.60	48.80	50.00	51.20	52.40	53.60	54.80
46"	49.00	50.20	51.40	52.60	53.80	55.00	56.20	57.40	58.60
48"	53.00	54.20	55.40	56.60	57.80	59.00	60.20	61.40	62.60
50"	57.20	58.50	59.70	60.90	62.10	63.30	64.50	65.70	66.90

WEIGHTS OF GALVANIZED STEEL RECTANGULAR ELBOWS *(cont.)*

Pounds per Elbow

Size	22"	24"	26"	28"	30"	32"	34"	36"	38"
22"	19.6	21.9	24.1	26.3	28.5	37.0	40.8	44.5	48.3
24"	21.9	24.2	26.4	28.6	30.9	39.5	43.6	47.2	50.8
26"	24.1	26.4	28.7	30.9	33.2	42.0	46.4	49.8	53.3
28"	26.3	28.6	30.9	33.3	35.6	44.5	49.2	52.5	55.8
30"	28.5	30.9	33.2	35.6	37.9	47.0	52.0	55.1	58.3
32"	37.0	39.5	42.0	44.5	47.0	49.5	52.0	57.8	60.8
34"	40.8	43.6	46.4	49.2	52.0	54.8	60.5	63.1	63.3
36"	44.5	47.2	49.8	52.5	55.1	57.8	60.5	63.1	65.8
38"	48.3	50.8	53.3	55.8	58.3	60.8	63.3	65.8	68.3
40"	52.1	54.6	57.1	59.6	62.1	64.6	67.1	69.6	72.1
42"	55.9	58.4	60.9	63.4	65.9	68.4	70.9	73.4	75.9
44"	59.6	62.1	64.6	67.1	69.6	72.1	74.6	77.1	79.6
46"	63.4	65.9	68.4	70.9	73.4	75.9	78.4	80.9	83.4

WEIGHTS OF GALVANIZED STEEL RECTANGULAR ELBOWS *(cont.)*

Pounds per Elbow

Size	22"	24"	26"	28"	30"	32"	34"	36"	38"
48"	67.2	69.7	72.2	74.7	77.2	79.7	82.2	84.7	87.2
50"	70.9	73.4	75.9	78.4	80.9	83.4	85.9	88.4	90.9
52"	74.8	77.3	79.8	82.3	84.8	87.3	89.8	92.3	94.7
54"	78.5	81.0	83.5	86.0	88.5	91.0	93.5	96.0	98.5
56"	97.2	100.0	103.0	106.0	109.0	112.0	115.0	118.0	121.0
58"	103.0	106.0	109.0	112.0	115.0	118.0	121.0	124.0	127.0
60"	110.0	113.0	116.0	119.0	122.0	125.0	128.0	131.0	134.0
62"	116.0	119.0	122.0	125.0	128.0	131.0	134.0	137.0	140.0
64"	122.0	125.0	129.0	132.0	135.0	138.0	141.0	144.0	147.0
66"	128.0	131.0	135.0	138.0	141.0	144.0	148.0	151.0	154.0
68"	135.0	138.0	142.0	145.0	148.0	151.0	154.0	157.0	160.0
70"	141.0	144.0	148.0	151.0	154.0	157.0	160.0	163.0	166.0
72"	147.0	150.0	154.0	157.0	160.0	163.0	166.0	169.0	172.0

WEIGHTS OF GALVANIZED STEEL RECTANGULAR ELBOWS (cont.)

Pounds per Elbow

Size	40"	42"	44"	46"	48"	50"	52"	54"	56"
40"	74.6	78.4	82.1	85.9	89.7	93.4	97.2	102.0	125.0
42"	78.4	82.3	86.2	90.1	94.0	97.9	102.0	106.0	130.0
44"	82.1	86.2	90.3	94.4	98.5	102.0	106.0	110.0	135.0
46"	85.9	90.1	94.4	98.7	102.0	106.0	110.0	113.0	140.0
48"	89.7	94.0	98.5	102.0	106.0	110.0	113.0	117.0	144.0
50"	93.4	97.9	102.0	106.0	110.0	113.0	117.0	122.0	149.0
52"	97.2	102.0	106.0	110.0	113.0	117.0	122.0	126.0	154.0
54"	102.0	106.0	110.0	113.0	117.0	122.0	126.0	131.0	159.0
56"	126.0	130.0	134.0	139.0	145.0	150.0	155.0	160.0	165.0
58"	132.0	137.0	142.0	147.0	151.0	156.0	161.0	166.0	170.0
60"	137.0	142.0	147.0	152.0	157.0	162.0	167.0	172.0	176.0
62"	143.0	148.0	153.0	158.0	163.0	168.0	173.0	178.0	182.0
64"	150.0	155.0	160.0	165.0	169.0	174.0	179.0	184.0	188.0
66"	156.0	161.0	166.0	171.0	176.0	181.0	186.0	191.0	195.0
68"	163.0	168.0	173.0	178.0	182.0	187.0	192.0	197.0	201.0
70"	170.0	175.0	180.0	185.0	189.0	194.0	199.0	204.0	208.0
72"	176.0	181.0	185.0	190.0	194.0	199.0	204.0	210.0	215.0

WEIGHTS OF GALVANIZED STEEL RECTANGULAR ELBOWS (cont.)

Pounds per Elbow

Size	58"	60"	62"	64"	66"	68"	70"	72"
58"	176	182	187	194	200	207	214	221
60"	182	187	194	200	207	214	221	227
62"	187	194	200	207	214	221	227	233
64"	194	200	207	214	221	227	233	239
66"	200	207	214	221	227	233	239	246
68"	207	214	221	227	233	239	246	252
70"	214	221	227	233	239	246	252	259
72"	221	227	233	239	246	252	259	266

WEIGHTS OF GALVANIZED STEEL RECTANGULAR TEES

Pounds per Tee

Size	6"	8"	10"	12"	16"	20"	24"	28"	32"
6"	1.81	2.11	2.42	2.72	4.24	5.01	5.78	6.55	8.90
8"	2.11	2.42	2.72	3.02	4.62	5.40	6.17	6.94	9.37
10"	2.42	2.72	3.02	3.32	5.01	5.78	6.55	7.32	9.84
12"	2.72	3.02	3.32	3.62	5.40	6.17	6.94	7.71	10.30
16"	4.24	4.62	5.01	5.40	6.17	6.94	7.71	8.48	11.30
20"	5.01	5.40	5.78	6.17	6.94	7.71	8.48	9.25	12.20
24"	5.78	6.17	6.55	6.94	7.71	8.48	9.25	10.00	13.10
28"	6.55	6.94	7.32	7.71	8.48	9.25	10.00	10.80	14.10
32"	8.90	9.37	9.84	10.30	11.30	12.20	13.10	14.10	15.00
36"	9.84	10.30	10.80	11.30	12.20	13.10	14.10	15.00	15.90
40"	10.80	11.30	11.70	12.20	13.10	14.10	15.00	15.90	16.90
44"	11.70	12.20	12.70	13.10	14.10	15.00	15.90	16.90	17.80
48"	12.70	13.10	13.60	14.10	15.00	15.90	16.90	17.80	18.70
52"	13.60	14.10	14.50	15.00	15.90	16.90	17.80	18.70	19.70
56"	16.00	17.60	18.20	18.80	19.60	21.00	22.10	23.20	24.30
60"	18.70	19.00	19.30	19.90	21.00	22.10	23.20	24.30	25.40

ESTIMATING GALVANIZED STEEL SPIRAL DUCTWORK

Estimating Math

Duct Perimeter = 3.1416 × Duct Diameter

Duct Total Area = Duct Perimeter × Duct Length

Duct Weight = Duct Weight per Sq. Ft. × Duct Total Area

Estimating Example

Find the weight for 26 gage 14" galvanized spiral ductwork 150' long

Calculation:

Duct diameter 14" /12 = 1.17 ft.

Duct perimeter is: 3.1416 × 1.17 = 3.68 ft.

Total duct area is: 3.68 × 150 = 552 sq. ft.

Add 25% waste for bracing, hangers, waste, and seams

Area = 552 × (1 + 25%) = 690 sq. ft.

Weight = 690 sq. ft. × 0.906 lbs./sq. ft. = 625 lbs.

WEIGHTS AND AREAS OF GALVANIZED STEEL SPIRAL DUCTS

Note: Includes 25% waste for bracing, hangers, waste, and seams

Diameter (Inches)	Sq. Ft. per Lin. Ft.	Lbs. per Lin. Ft.					
		26 Gauge	24 Gauge	22 Gauge	20 Gauge	18 Gauge
4	1.05	1.19	1.51	1.84	2.17	2.82
5	1.31	1.48	1.89	2.30	2.71	3.53
6	1.57	1.78	2.27	2.76	3.25	4.23
7	1.83	2.08	2.65	3.22	3.79	4.94
8	2.09	2.37	3.03	3.68	4.34	5.64
9	2.36	2.67	3.40	4.14	4.88	6.35
10	2.62	2.96	3.78	4.60	5.42	7.06
11	2.88	3.26	4.16	5.06	5.96	7.76
12	3.14	3.56	4.54	5.52	6.50	8.47
13	3.40	3.85	4.92	5.98	7.05	9.17
14	3.67	4.15	5.30	6.44	7.59	9.88
15	3.93	4.45	5.67	6.90	8.13	10.58
16	4.19	4.74	6.05	7.36	8.67	11.29

WEIGHTS AND AREAS OF GALVANIZED STEEL SPIRAL DUCTS (cont.)

Note: Includes 25% waste for bracing, hangers, waste, and seams

Diameter (Inches)	Sq. Ft. per Lin. Ft.	Lbs. per Lin. Ft.					
		26 Gauge	24 Gauge	22 Gauge	20 Gauge	18 Gauge
17	4.45	5.04	6.43	7.82	9.21	11.99
18	4.71	5.34	6.81	8.28	9.75	12.70
19	4.97	5.63	7.19	8.74	10.30	13.41
20	5.24	5.93	7.57	9.20	10.84	14.11
21	5.50	6.23	7.94	9.66	11.38	14.82
22	5.76	6.52	8.32	10.12	11.92	15.52
23	6.02	6.82	8.70	10.58	12.46	16.23
24	6.28	7.12	9.08	11.04	13.01	16.93
25	6.55	7.41	9.46	11.50	13.55	17.64
26	6.81	7.71	9.84	11.96	14.09	18.34
27	7.07	8.01	10.21	12.42	14.63	19.05
28	7.33	8.30	10.59	12.88	15.17	19.76

WEIGHTS AND AREAS OF GALVANIZED STEEL SPIRAL DUCTS (cont.)

Note: Includes 25% waste for bracing, hangers, waste, and seams

Diameter (Inches)	Sq. Ft. per Lin. Ft.	Lbs. per Lin. Ft.					
		26 Gauge	24 Gauge	22 Gauge	20 Gauge	18 Gauge
29	7.59	8.60	10.97	13.34	15.72	20.46
30	7.85	8.89	11.35	13.80	16.26	21.17
31	8.12	9.19	11.73	14.26	16.80	21.87
32	8.38	9.49	12.11	14.72	17.34	22.58
33	8.64	9.78	12.48	15.18	17.88	23.28
34	8.90	10.08	12.86	15.64	18.43	23.99
35	9.16	10.38	13.24	16.10	18.97	24.69
36	9.42	10.67	13.62	16.56	19.51	25.40
37	9.69	10.97	14.00	17.02	20.05	26.11
38	9.95	11.27	14.38	17.48	20.59	26.81
39	10.21	11.56	14.75	17.94	21.14	27.52
40	10.47	11.86	15.13	18.40	18.40	28.22

WEIGHTS OF GALVANIZED STEEL ROUND SPIRAL ELBOWS

	Pounds per Elbow	
Diameter	90 Degree Elbow	45 Degree Elbow
4"	2.2	1.3
5"	3.3	1.9
6"	4.3	2.5
7"	5.8	3.3
8"	7.3	4.3
9"	8.8	5.3
10"	11.8	7.5
12"	16.3	10.0
14"	22.0	13.0
16"	28.3	15.8
18"	34.5	19.0
20"	41.5	23.5
22"	48.3	27.5
24"	57.5	32.0
26"	68.8	37.5
28"	76.8	42.8
30"	87.0	48.0
32"	99.5	55.0
34"	112.0	61.0
36"	161.0	89.5

WEIGHTS OF GALVANIZED STEEL
ROUND SPIRAL COUPLING AND REDUCER

Pounds per Fitting

Diameter	Coupling	Reducer
4"	0.6	1.2
5"	0.7	1.4
6"	0.9	1.8
7"	1.0	2.0
8"	1.2	2.4
9"	2.6	3.3
10"	2.9	4.4
12"	3.5	5.3
14"	4.1	6.2
16"	4.6	6.9
18"	5.2	7.8
20"	5.8	8.7
22"	6.4	9.6
24"	6.9	10.3
26"	7.5	11.2
28"	8.1	12.1
30"	8.7	13.0
32"	9.2	13.8
34"	9.8	14.7
36"	10.4	15.6

WEIGHTS OF GALVANIZED STEEL ROUND SPIRAL TEES WITH REDUCING RUN AND BRANCH

Pounds per Tee

Largest Run Diameter	Branch Diameter													
	3"	4"	5"	6"	7"	8"	9"	10"	12"	14"	16"	18"	20"	22"
4"	2.3													
5"	2.8	3.0												
6"	3.2	3.5	3.7											
7"	3.7	3.9	4.1	4.5										
8"	4.1	4.4	4.7	5.0	5.3									
9"	4.5	4.8	5.1	5.5	5.8	6.2								
10"	6.4	6.7	7.1	7.5	7.8	8.2	8.6							
12"	7.7	8.1	8.5	8.9	9.4	9.9	10.3	10.7						
14"	8.9	9.4	10.0	10.5	11.0	11.5	12.1	12.6	13.6					
16"	10.2	10.8	11.4	12.0	12.5	13.1	13.7	14.3	15.5	16.7				

WEIGHTS OF GALVANIZED STEEL ROUND SPIRAL TEES WITH REDUCING RUN AND BRANCH *(cont.)*

Pounds per Tee

Largest Run Diameter	Branch Diameter												
	3"	4"	5"	6"	7"	8"	9"	10"	12"	14"	16"	18"	20"	22"
18"	11.5	12.1	12.7	13.5	14.1	14.8	15.5	16.1	17.5	18.8	19.2			
20"	12.7	13.4	14.2	14.9	17.7	16.4	17.2	17.9	19.4	20.9	22.4	23.9		
22"	14.0	14.8	15.6	16.4	17.2	18.1	18.9	19.7	21.3	23.0	24.6	26.2	27.9	
24"	15.3	16.1	17.0	17.8	18.7	19.6	20.5	21.4	23.2	25.0	26.8	28.5	30.3	32.1
26"	16.5	17.5	18.5	19.5	20.4	21.3	22.3	23.3	25.2	26.2	29.1	31.1	33.0	34.9
28"	17.7	18.7	19.8	20.8	21.9	22.9	23.9	25.0	27.0	29.2	31.2	33.2	35.4	37.4
30"	18.9	20.0	21.1	22.2	23.3	24.5	25.5	26.7	28.9	31.1	33.3	35.5	37.7	40.0
32"	20.4	21.6	22.6	24.0	25.2	26.4	27.6	28.8	31.2	33.6	36.0	38.4	40.8	42.2
34"	21.6	22.9	24.2	25.4	26.7	28.0	29.3	30.5	33.1	35.6	38.1	40.7	42.2	45.6
36"	29.6	31.4	33.1	34.8	36.6	38.6	40.1	41.8	45.3	48.8	52.2	55.7	59.2	62.7

7-134

WEIGHTS OF GALVANIZED STEEL ROUND SPIRAL CROSS WITH REDUCING BRANCH

Pounds per Tee

Largest Run Diameter	Branch Diameter													
	3"	4"	5"	6"	7"	8"	9"	10"	12"	14"	16"	18"	20"	22"
4"	2.6	–	–	–	–	–	–	–	–	–	–	–	–	–
5"	3.2	3.6	–	–	–	–	–	–	–	–	–	–	–	–
6"	3.8	4.2	4.6	–	–	–	–	–	–	–	–	–	–	–
7"	4.4	4.9	5.4	5.8	–	–	–	–	–	–	–	–	–	–
8"	4.9	5.4	6.0	6.5	7.0	–	–	–	–	–	–	–	–	–
9"	5.4	6.0	6.6	7.2	7.7	8.3	–	–	–	–	–	–	–	–
10"	7.6	8.2	8.9	9.5	10.2	10.8	11.4	–	–	–	–	–	–	–
12"	8.8	9.6	10.3	11.0	11.7	12.5	13.2	13.8	–	–	–	–	–	–
14"	10.1	10.9	11.7	12.6	13.4	14.2	15.1	15.9	17.5	–	–	–	–	–
16"	11.3	12.2	13.1	14.0	15.0	15.9	16.8	17.7	19.5	21.4	–	–	–	–

WEIGHTS OF GALVANIZED STEEL ROUND SPIRAL CROSS WITH REDUCING BRANCH *(cont.)*

Pounds per Tee

Largest Run Diameter	Branch Diameter																			
	3"	4"	5"	6"	7"	8"	9"	10"	12"	14"	16"	18"	20"	22"
18"	12.5	13.5	14.5	15.5	16.5	17.5	18.5	19.5	21.5	23.5	25.5	—	—	—
20"	13.8	15.0	16.1	17.2	18.2	19.4	20.6	21.7	23.9	26.2	28.4	30.6	—	—
22"	15.2	16.4	17.7	18.9	20.1	21.4	22.6	23.8	26.2	28.7	31.2	33.6	36.1	—
24"	16.6	17.9	19.3	20.6	22.0	23.3	24.6	26.0	28.7	31.5	34.1	36.7	39.4	42.1
26"	18.0	19.4	20.9	22.3	23.8	25.2	26.7	28.2	31.1	33.9	36.9	39.8	42.7	45.6
28"	19.3	20.8	22.4	24.0	25.5	27.1	28.6	30.2	32.3	36.4	39.6	42.7	45.8	48.9
30"	20.6	22.3	23.9	25.6	27.3	28.9	30.6	32.3	35.6	38.9	42.3	45.6	49.9	52.2
32"	22.2	24.0	25.8	27.6	29.4	31.2	33.0	34.8	38.4	42.0	45.6	49.1	52.8	56.5
34"	23.5	25.4	27.3	29.2	31.1	33.1	34.9	36.9	40.7	44.5	48.3	52.1	55.9	59.8
36"	32.2	34.8	37.4	40.0	42.8	45.3	48.0	50.5	55.7	61.0	66.2	71.5	76.7	82.0

ESTIMATING ALUMINUM RECTANGULAR DUCTWORK

Ductwork Gauge	Weight (lb./sq. ft.)	Section Long Side Max Dimension (in.)
22	0.357	30"
20	0.450	54"
18	0.568	84"
16	0.716	85" and up

Estimating Example

Find the weight for 36" × 15" aluminum rectangular ductwork 20' long

Calculation:

Section long side dimension is 36", more than 30" but less than 54"

So this is a 20 gauge ductwork

Area = (36 + 15) × 2 /12 × 20 = 170 sq. ft.

Add 25% waste for bracing, hangers, waste, and seams

Area = 170 × (1 + 25%) = 213 sq. ft.

Weight = 213 sq. ft. × 0.450 lbs./sq. ft. = 96 lbs.

WEIGHTS AND AREAS OF ALUMINUM RECTANGULAR DUCTS

Note: Includes 25% waste for bracing, hangers, waste, and seams

Width + Depth (Inches)	Sq. Ft. per Lin. Ft.	Lbs. per Lin. Ft.			
		22 Gauge	20 Gauge	18 Gauge	16 Gauge
22	3.67	1.64	2.06	N/A	N/A
23	3.83	1.71	2.16	N/A	N/A
24	4.00	1.79	2.25	N/A	N/A
25	4.17	1.86	2.34	N/A	N/A
26	4.33	1.93	2.44	N/A	N/A
27	4.50	2.01	2.53	N/A	N/A
28	4.67	2.08	2.63	N/A	N/A
29	4.83	2.16	2.72	N/A	N/A
30	5.00	2.23	2.81	N/A	N/A
31	5.17	2.31	2.91	N/A	N/A
32	5.33	2.38	3.00	N/A	N/A
33	5.50	2.45	3.09	N/A	N/A
34	5.67	2.53	3.19	N/A	N/A

WEIGHTS AND AREAS OF ALUMINUM RECTANGULAR DUCTS (cont.)

Note: Includes 25% waste for bracing, hangers, waste, and seams

Width + Depth (Inches)	Sq. Ft. per Lin. Ft.	Lbs. per Lin. Ft.			
		22 Gauge	20 Gauge	18 Gauge	16 Gauge
35	5.83	2.60	3.28	N/A	N/A
36	6.00	2.68	3.38	N/A	N/A
37	6.17	2.75	3.47	N/A	N/A
38	6.33	2.83	3.56	N/A	N/A
39	6.50	2.90	3.66	N/A	N/A
40	6.67	2.98	3.75	4.73	N/A
41	6.83	3.05	3.84	4.85	N/A
42	7.00	3.12	3.94	4.97	N/A
43	7.17	3.20	4.03	5.09	N/A
44	7.33	3.27	4.13	5.21	N/A
45	7.50	3.35	4.22	5.33	N/A
46	7.67	3.42	4.31	5.44	N/A
47	7.83	3.50	4.41	5.56	N/A

WEIGHTS AND AREAS OF ALUMINUM RECTANGULAR DUCTS (cont.)

Note: Includes 25% waste for bracing, hangers, waste, and seams

Width + Depth (Inches)	Sq. Ft. per Lin. Ft.	Lbs. per Lin. Ft.			
		22 Gauge	20 Gauge	18 Gauge	16 Gauge
48	8.00	3.57	4.50	5.68	N/A
49	8.17	3.64	4.59	5.80	N/A
50	8.33	3.72	4.69	5.92	N/A
51	8.50	3.79	4.78	6.04	N/A
52	8.67	3.87	4.88	6.15	N/A
53	8.83	3.94	4.97	6.27	N/A
54	9.00	4.02	5.06	6.39	N/A
55	9.17	4.09	5.16	6.51	N/A
56	9.33	4.17	5.25	6.63	N/A
57	9.50	4.24	5.34	6.75	N/A
58	9.67	4.31	5.44	6.86	N/A
59	9.83	4.39	5.53	6.98	N/A
60	10.00	4.46	5.63	7.10	N/A

WEIGHTS AND AREAS OF ALUMINUM RECTANGULAR DUCTS (cont.)

Note: Includes 25% waste for bracing, hangers, waste, and seams

Width + Depth (Inches)	Sq. Ft. per Lin. Ft.	Lbs. per Lin. Ft.			
		22 Gauge	20 Gauge	18 Gauge	16 Gauge
61	10.17	N/A	5.72	7.22	N/A
62	10.33	N/A	5.81	7.34	N/A
63	10.50	N/A	5.91	7.46	N/A
64	10.67	N/A	6.00	7.57	N/A
65	10.83	N/A	6.09	7.69	N/A
66	11.00	N/A	6.19	7.81	N/A
67	11.17	N/A	6.28	7.93	N/A
68	11.33	N/A	6.38	8.05	N/A
69	11.50	N/A	6.47	8.17	N/A
70	11.67	N/A	6.56	8.28	N/A
71	11.83	N/A	6.66	8.40	N/A
72	12.00	N/A	6.75	8.52	10.74
73	12.17	N/A	6.84	8.64	10.89

WEIGHTS AND AREAS OF ALUMINUM RECTANGULAR DUCTS (cont.)

Note: Includes 25% waste for bracing, hangers, waste, and seams

Width + Depth (Inches)	Sq. Ft. per Lin. Ft.	Lbs. per Lin. Ft.			
		22 Gauge	20 Gauge	18 Gauge	16 Gauge
74	12.33	N/A	6.94	8.76	11.04
75	12.50	N/A	7.03	8.88	11.19
76	12.67	N/A	7.13	8.99	11.34
77	12.83	N/A	7.22	9.11	11.49
78	13.00	N/A	7.31	9.23	11.64
79	13.17	N/A	7.41	9.35	11.78
80	13.33	N/A	7.50	9.47	11.93
81	13.50	N/A	7.59	9.59	12.08
82	13.67	N/A	7.69	9.70	12.23
83	13.83	N/A	7.78	9.82	12.38
84	14.00	N/A	7.88	9.94	12.53
85	14.17	N/A	7.97	10.06	12.68
86	14.33	N/A	8.06	10.18	12.83

WEIGHTS AND AREAS OF ALUMINUM RECTANGULAR DUCTS (cont.)

Note: Includes 25% waste for bracing, hangers, waste, and seams

Width + Depth (Inches)	Sq. Ft. per Lin. Ft.	Lbs. per Lin. Ft.			
		22 Gauge	20 Gauge	18 Gauge	16 Gauge
87	14.50	N/A	8.16	10.30	12.98
88	14.67	N/A	8.25	10.41	13.13
89	14.83	N/A	8.34	10.53	13.28
90	15.00	N/A	8.44	10.65	13.43
91	15.17	N/A	8.53	10.77	13.57
92	15.33	N/A	8.63	10.89	13.72
93	15.50	N/A	8.72	11.01	13.87
94	15.67	N/A	8.81	11.12	14.02
95	15.83	N/A	8.91	11.24	14.17
96	16.00	N/A	9.00	11.36	14.32
97	16.17	N/A	9.09	11.48	14.47
98	16.33	N/A	9.19	11.60	14.62
99	16.50	N/A	9.28	11.72	14.77

WEIGHTS AND AREAS OF ALUMINUM RECTANGULAR DUCTS (cont.)

Note: Includes 25% waste for bracing, hangers, waste, and seams

Width + Depth (Inches)	Sq. Ft. per Lin. Ft.	Lbs. per Lin. Ft.			
		22 Gauge	20 Gauge	18 Gauge	16 Gauge
100	16.67	N/A	9.38	11.83	14.92
101	16.83	N/A	9.47	11.95	15.07
102	17.00	N/A	9.56	12.07	15.22
103	17.17	N/A	9.66	12.19	15.36
104	17.33	N/A	9.75	12.31	15.51
105	17.50	N/A	9.84	12.43	15.66
106	17.67	N/A	9.94	12.54	15.81
107	17.83	N/A	10.03	12.66	15.96
108	18.00	N/A	10.13	12.78	16.11
109	18.17	N/A	10.22	12.90	16.26
110	18.33	N/A	10.31	13.02	16.41
111	18.50	N/A	10.41	13.14	16.56
112	18.67	N/A	10.50	13.25	16.71

WEIGHTS AND AREAS OF ALUMINUM RECTANGULAR DUCTS (cont.)

Note: Includes 25% waste for bracing, hangers, waste, and seams

Width + Depth (Inches)	Sq. Ft. per Lin. Ft.	Lbs. per Lin. Ft.			
		22 Gauge	20 Gauge	18 Gauge	16 Gauge
113	18.83	N/A	10.59	13.37	16.86
114	19.00	N/A	10.69	13.49	17.01
115	19.17	N/A	N/A	13.61	17.15
116	19.33	N/A	N/A	13.73	17.30
117	19.50	N/A	N/A	13.85	17.45
118	19.67	N/A	N/A	13.96	17.60
119	19.83	N/A	N/A	14.08	17.75
120	20.00	N/A	N/A	14.20	17.90
122	20.33	N/A	N/A	14.44	18.20
124	20.67	N/A	N/A	14.67	18.50
126	21.00	N/A	N/A	14.91	18.80
128	21.33	N/A	N/A	15.15	19.09
130	21.67	N/A	N/A	15.38	19.39

WEIGHTS AND AREAS OF ALUMINUM RECTANGULAR DUCTS *(cont.)*

Note: Includes 25% waste for bracing, hangers, waste, and seams

Width + Depth (Inches)	Sq. Ft. per Lin. Ft.	Lbs. per Lin. Ft.				
		22 Gauge	20 Gauge	18 Gauge	16 Gauge	
132	22.00	N/A	N/A	15.62	19.69	
134	22.33	N/A	N/A	15.86	19.99	
136	22.67	N/A	N/A	16.09	20.29	
138	23.00	N/A	N/A	16.33	20.59	
140	23.33	N/A	N/A	16.57	20.88	
142	23.67	N/A	N/A	16.80	21.18	
144	24.00	N/A	N/A	17.04	21.48	
146	24.33	N/A	N/A	17.28	21.78	
148	24.67	N/A	N/A	17.51	22.08	
150	25.00	N/A	N/A	17.75	22.38	
152	25.33	N/A	N/A	17.99	22.67	
154	25.67	N/A	N/A	18.22	22.97	
156	26.00	N/A	N/A	18.46	23.27	

WEIGHTS AND AREAS OF ALUMINUM RECTANGULAR DUCTS *(cont.)*

Note: Includes 25% waste for bracing, hangers, waste, and seams

Width + Depth (Inches)	Sq. Ft. per Lin. Ft.	Lbs. per Lin. Ft.				
		22 Gauge	20 Gauge	18 Gauge	16 Gauge	
158	26.33	N/A	N/A	18.70	23.57
160	26.67	N/A	N/A	18.93	23.87
162	27.00	N/A	N/A	19.17	24.17
164	27.33	N/A	N/A	19.41	24.46
166	27.67	N/A	N/A	19.64	24.76
168	28.00	N/A	N/A	19.88	25.06
170	28.33	N/A	N/A	20.12	25.36
172	28.67	N/A	N/A	20.35	25.66
174	29.00	N/A	N/A	20.59	25.96
176	29.33	N/A	N/A	20.83	26.25
178	29.67	N/A	N/A	21.06	26.55
180	30.00	N/A	N/A	21.30	26.85

ESTIMATING EQUIPMENT PAD

Concrete Volume per Sq. Ft. of Slab

Thickness (in.)	Volume (cy.)
4"	0.014
6"	0.021
8"	0.028
10"	0.035
12"	0.043
14"	0.050
16"	0.057
18"	0.064
20"	0.071
24"	0.085
30"	0.106

Estimating Example:

For an equipment pad of 12' long, 10' wide, 8" thick

Area of the pad: $10 \times 12 = 120$ sq. ft.

Concrete needed is about $120 \times 0.028 = 3.36$ cy.

ESTIMATING SOLDER, FLUX, AND GAS FOR SOLDERING 100 COPPER JOINTS

Solder is estimated using the table below:

Pipe & Fitting (Inches)	Solder (Pounds)
3/8"	1.1
1/2"	1.5
3/4"	1.8
1"	2.3
1 1/4"	3.4
1 1/2"	3.9
2"	4.6
2 1/2"	5.4
3"	6.9
3 1/2"	7.5
4"	8.5
5"	11.0
6"	15.0
8"	32.0
10"	42.0

Flux is estimated at two oz. for each pound of solder.
Gas is about 1 tank per every 500 joints or for each 3 days of continuous soldering.

Estimating Example:

For 1000 ea. 6"copper joints
Solder: $1000/100 \times 15.0 = 150$ lbs.
Flux: $150 \times 2 = 300$ oz.
Gas: $1000/500 = 2$ tanks

TANK VOLUME (ROUND) PER FOOT OF DEPTH

Round Tank Diameter	Volume (U.S. Gallons) per Foot of Depth
1' 0"	5.88
1' 2"	8.04
1' 4"	10.39
1' 6"	13.22
1' 8"	16.39
1' 10"	19.68
2' 0"	23.50
2' 2"	27.67
2' 4"	31.90
2' 6"	36.72
2' 8"	41.89
2' 10"	47.06
3' 0"	52.88
3' 2"	59.04
3' 4"	65.15
3' 6"	71.98
3' 8"	79.14
3' 10"	86.19
4' 0"	94.01
4' 2"	102.17
4' 4"	110.16
4' 6"	118.98
4' 8"	128.14
4' 10"	137.07
5' 0"	146.89
5' 2"	157.05
5' 4"	166.92
5' 6"	177.74
5' 8"	188.89
5' 10"	199.70
6' 0"	211.52
6' 6"	248.24
7' 0"	287.90
7' 6"	330.50

TANK VOLUME (RECTANGULAR) PER FOOT OF DEPTH

Tank Dimension	Volume (U.S. Gallons) per Foot of Depth
2' × 2'	29.92
2' × 2' 6"	37.41
2' × 3'	44.89
2' × 3' 6"	52.37
2' × 4'	59.85
2' × 4' 6"	67.33
2' × 5'	74.81
2' 6" × 2' 6"	46.76
2' 6" × 3'	56.11
2' 6" × 3' 6"	65.46
2' 6" × 4'	74.81
2' 6" × 4' 6"	84.16
2' 6" × 5'	93.51
3' × 3'	67.33
3' × 3' 6"	78.55
3' × 4'	89.77
3' × 4' 6"	100.99
3' × 5'	112.22
4' × 4'	119.70
4' × 4' 6"	134.66
4' × 5'	149.62
4' 6" × 4' 6"	151.49
4' 6" × 5'	168.32
5' × 5'	187.03

ESTIMATING EQUIPMENT PAD

Concrete Volume per Sq. Ft. of Slab

Thickness (in.)	Volume (cy)
4"	0.014
6"	0.021
8"	0.028
10"	0.035
12"	0.043
14"	0.050
16"	0.057
18"	0.064
20"	0.071
24"	0.085
30"	0.106

Estimating Example:

For an equipment pad of 12' long, 10' wide, 8" thick

Area of the pad: $10 \times 12 = 120$ SF

Concrete needed is about $120 \times 0.028 = 3.36$ CY

CHAPTER 8
Abbreviations

A	Area Square Feet, Ampere	BSTR	Booster
AC	Alternating Current, Air Conditioning	BTM	Bottom
		BTU	British Thermal Unit
ACC	Accessory	C	Celsius
ACCU	Accumulator	CAV	Constant Air Volume
ACPTR	Acceptor	CB	Circuit Breaker
ACT	Actuator	CCW	Counter Clock Wise
ACTI	Activated	CFM	Cubic Feet per Minute
ADD'L	Additional		
ADJ	Adjustable	CHLL	Chiller
ADPT	Adapter	CHRM	Chrome
AL	Aluminum	CHW	Chilled Water
ALRM	Alarm	CI	Cast Iron
ALT	Altitude, Alternate	CIR	Circuit
ANNN	Annunciator	CIRC	Circulator
AP	Accessory Package	CNDCT	Conductor
ASHRAE	American Society of Heating Refrigeration and Air Conditioning Engineers	CNDT	Conduit
		COMP	Compressor
		COND	Condenser
		CONT	Contactor
		COUP	Coupling
ASME	American Society of Mechanical Engineers.	CU	Copper, Cubic
		CW	Clockwise
		CWT	100 Pounds
ASTM	American Society of Testing and Materials	DBL	Double
		DC	Direct Current
ATNT	Attenuator	DDC	Direct Digital Control
BAL	Balancing	DECO	Decorative
BFFL	Baffle	DEV	Device
BHP	Boiler Horsepower	DFFSR	Diffuser
BLCK	Block	DFRST	Defrost
BPM	Blows per Minute	DHW	Domestic Hot Water
BRD	Board	DIA	Diaphragm
BRG	Bearing	DIAG	Diagram
BRKR	Breaker	DIM	Dimming

ABBREVIATIONS *(cont.)*

DISC	Disconnect, Discharge	FRSHN	Freshener
DISP	Disposable, Display	FRZR	Freezer
DIST	Distribution	FURN	Furnace
DMP	Damper	Fy	Minimum Yield Stress of Steel
DRFT	Draft	G	Gram
DWV	Drain Waste Vent	GA	Gauge
EA	Each	GAL	Gallon
ECON	Economizer	GALV	Galvanized
EFF	Efficiency	GENR	Generator
ELEC	Electric	GPH	Gallons per hour
ELEM	Element	GPM	Gallons per minute
ELIM	Eliminator	GRAV	Gravity
EMER	Emergency	GRD	Guard
ENCL	Enclosure	GRN	Green
ENER	Energy	GRND	Ground
EQL	Equalizer	GSKT	Gasket
ES	Energy Saver	H/C	Heat/Cool
ESCT	Escutcheon	HDWR	Hardware
EVAP	Evaporator	HE	High Efficiency
EW	Each Way	HLDBK	Holdback
EXC	Exchanger	HLF	Half
EXH	Exhauster	HLW	Hollow
EXP	Expansion, Exposure	HMD	Humidistat
EXT	Exterior, Extension, Extrusion, External	HNDL	Handle
		HNG	Hinge
F	Fahrenheit, Fan	HNGR	Hanger
FACT	Factory	HORZ	Horizontal
FC	Foot-candles	HOUS	Housing
FEM	Female	HP	Horsepower, High Pressure
FG	Fiberglass		
FHLDR	Fuseholder	HTG	Heating
FILT	Filter	HTR	Heater
FIT	Fitting	HV	High Voltage
FLNG	Flange	HZ	Hertz
FLRS	Fluorescent	ID	Indoor, Inside Diameter/Dimension
FM	Factory Mutual		
FNNL	Funnel	IGN	Ignition
FPM	Feet per Minute	IMP	Impeller
FR	Fire-rated	IND	Indicator, Inducer
FRCD	Forced	INI	Initiation, Injection
FRM	Frame	INLN	Inline

ABBREVIATIONS *(cont.)*

INLT	Inlet	**MAN**	Manual
INNR	Inner	**MANI**	Manifold
INSP	Inspection	**MBH**	Thousand BTU
INSRT	Insert		per Hour
INST	Installation	**MDFR**	Modifier
INT	Interlock, Interrupter,	**MFFLR**	Muffler
	Internal, Intermediate	**MG**	Milligram
INTG	Integrated	**MH**	Man-Hour, Manhole
INTK	Intake	**MHZ**	Megahertz
INTR	Intermittant	**MM**	Millimeter
INTRF	Interface	**MN**	Main
INVRTR	Inverter	**MNT**	Mount
IPS	Iron Pipe Size	**MNTR**	Monitor
J	Joule	**MOD**	Module, Modulator
JMPR	Jumper	**MTR**	Motor
JNCT	Junction	**MTRL**	Material
JNT	Joint	**MV**	Milivolt
K	Thousand, Kelvin	**MW**	Megawatt
KG	Kilogram	**NA**	Not Applicable,
KIP	1000 Pounds		Not Available
KNB	Knob	**N/C**	Normally closed
KNRL	Knurled	**N/O**	Normally open
KVA	Kilovolt Ampere	**NEG**	Negative
KYPD	Keypad	**NEUT**	Neutralizer
LEV	Level	**NG**	Natural gas
LG	Large, Long, Length	**NLA**	No Longer Available
LH	Left Hand	**NLB**	Non-Load-Bearing
LIQ	Liquid	**NOC**	Not Otherwise
LL	Live Load		Classified
LMT	Limit	**NON**	Non (without)
LNR	Liner	**NP**	Not Protected
LP	Low Pressure, Liquid	**NPPL**	Nipple
	propane gas	**NUM**	Numerical
LS	Lump Sum	**NZZL**	Nozzle
LTCH	Latch	**OB**	Outboard
LTHR	Leather	**OBS**	Obsolete
LUB	Lubricant	**OD**	Outdoor, Outside
LV	Low Voltage		Diameter/Dimension
LVL	Leveler	**OFST**	Offset
LVR	Louver	**OH**	Overhead
M	Thousand, Material	**O & P**	Overhead & Profit
MAG	Magnetic	**OPP**	Opposite

ABBREVIATIONS (cont.)

OPTM	Optimizer	RDUN	Redundant
ORG	Original	RECEPT	Receptacle
OS	Oversize	REF	Reference
OVL	Oval	REFR	Refrigerator
OVRLD	Overload	REG	Regulator
PERF	Perforated	REM	Remote
PERM	Permanent	REPL	Replace
PIST	Piston	RES	Resilient
PLG	Plug	REV	Reversing
PLNGR	Plunger	RFLC	Reflector
PLNM	Plenum	RFRG	Refrigeration
PLR	Polarized	RH	Right Hand
PLT	Pilot	RHW	Residential Hot
PLY	Pulley		Water, Rubber/Heat/
PMP	Pump		Water Resistant
PNL	Panel		
PNT	Paint, Point	RLF	Relief
POS	Position	RLLR	Roller
PPM	Parts per Million	RLY	Relay
PRI	Primary	RMP	Ramp
PRMR	Primer	RMT	Remote
PRP	Propane	RND	Round
PSF	Pounds per Square	RPB	Rapid Press Balance
	Foot	RPM	Revolutions per
PSI	Pounds per Square		Minute
	Inch	RPR	Repair
PSIG	Pounds per Square	RS	Rolled Steel, Rapid
	Inch Gauge		Start
P & T	Pressure &	RSR	Riser
	Temperature	RSRV	Reserve
PVT	Pivot	RST	Reset
PWR	Power	RSTR	Restrictor
QTY	Quantity	RTNR	Retainer
RADI	Radiation	RTR	Rater
RBBR	Rubber	RTRN	Return
RCKR	Rocker	RUB	Rubber
RCPT	Receptacle	RVT	Rivet
RCTR	Reactor	RWRK	Rework
RCVR	Receiver, Recovery	S4S	Surface 4 Sides
RCYC	Recycling	SCFM	Standard Cubic
RDCR	Reducer		Feet per Minute
RDNT	Radiant	SCH	Schedule
		SCKT	Socket

ABBREVIATIONS (cont.)

SCRN	Screen	**TRANS**	Transistor,
SDDL	Saddle		Transmission
SEC	Secondary	**TRBLTR**	Turbulator
SFCA	Square Foot Contact	**TRM**	Terminal
	Area	**TRNSD**	Transducer
SHD	Shade	**TRSN**	Torsion
SHLD	Shield	**TURB**	Turbulator
SHNT	Shunt	**UL**	Underwriters
SLNT	Sealant		Laboratory
SLP	Sulpher	**UNLD**	Unloader
SM	Small	**UPLS**	Upholstery
SMP	Sump	**URD**	Underground
SNK	Sink		Residential
SP	Static Pressure,		Distribution
	Single Pole,	**UTIL**	Utility
	Self-Propelled	**UV**	Ultraviolet
SPD	Speed	**VAC**	Vacuum
SPRT	Support	**VAR**	Variable
STAT	Status	**VAV**	Variable Air Volume
STBL	Stabilizer	**VEST**	Vestibule
STC	Sound Transmission	**VLV**	Valve
	Coefficient	**VNT**	Vent
STP	Standard	**VPR**	Vapor
	Tempressure	**VS**	Variable Speed
	and Pressure	**WE**	White Enamel
STRNR	Strainer	**WF**	Wide Flange
STRT	Start	**WHL**	Wheel
SUCT	Suction	**WR**	Water Resistant
SUPP	Supplemental, Supply	**WTR**	Water
SVC	Service	**XX-OHM**	Ohm Rating
SW	Switch		
SYS	System		
T & C	Threaded & Coupled		
TEMP	Temperature		
TFLN	Teflon		
TGGL	Toggle		
THRM	Thermometer,		
	Thermostat		
TMPL	Template		
TMPR	Tempering		
TNNL	Tunnel		

About The Author

Adam Ding is a professional estimator with extensive experience in a variety of commercial, residential, institutional, industrial, and infrastructure projects. He holds a Master's degree in Building Construction from Auburn University.

Adam has had a successful career in estimating projects of different sizes, ranging from large-scale cost planning to detailed trade take-off. Having bid countless construction jobs, he developed a unique and easy-to-understand estimating approach.

Adam taught computer classes in Auburn University and also owns the copyrights for a number of computer estimating programs. Currently a registered member in the Canadian Institute of Quantity Surveyors (CIQS), he continues to provide cost estimating services to building professionals in North America, Canada, and Asia Pacific regions.